STP 1151

Nondestructive Testing Standards—Present and Future

Harold Berger and Leonard Mordfin, editors

ASTM Publication Code Number (PCN)
04-011510-22

ASTM
1916 Race Street
Philadelphia, PA 19103

Library of Congress Cataloging in Publication Data

Nondestructive testing standards—present and future/Harold Berger
and Leonard Mordfin, editors.
    p.   cm. — (STP: 1151)
    "Papers presented at the Symposium on Nondestructive Testing
Standards II: New opportunities for Increased World Trade through
Accepted Standards for NDT and Quality, held 9-11 April 1991 in
Gaithersburg, Maryland"—Foreword.
    "Sponsored by ASTM Committee E-7 on Nondestructive Testing and the
National Institute of Standards and Technology (NIST) in cooperation
with the American Society for Nondestructive Testing and the
American Welding Society"—Foreword.
    Includes bibliographical references and index.
    ISBN 0-8031-1487-7
    1. Non-destructive testing—Standards—Congresses. I. Berger,
Harold. II. Mordfin, Leonard. III. American Society for Testing
and Materials. Committee E-7 on Nondestructive Testing.
    IV. Series: ASTM special technical publication: 1151.
TA417.2.N678   1992
620.1'127'021873—dc20   92-20433
                                                 CIP

Photocopy Rights

Peer Review Policy

Each paper published in this volume was evaluated by three peer reviewers. The authors
addressed all of the reviewers' comments to the satisfaction of both the technical editor(s) and the
ASTM Committee on Publications.

The qualtiy of the papers in this publication reflects not only the obvious efforts of the authors and
the technical editor(s), but also the work of these peer reviewers. The ASTM Committee on Publications
acknowledges with appreciation their dedication and contribution to time and effort on behalf of ASTM.

Printed in Baltimore, MD
August 1992

Foreword

This publication, *Nondestructive Testing Standards—Present and Future,* contains papers presented at the Symposium on Nondestructive Testing Standards II: New Opportunities for Increased World Trade Through Accepted Standards for NDT and Quality, held 9–11 April 1991 in Gaithersburg, Maryland. The symposium was sponsored by ASTM Committee E-7 on Nondestructive Testing and the National Institute of Standards and Technology (NIST) in cooperation with the American Society for Nondestructive Testing and the American Welding Society. James Borucki, Ardrox, Inc., served as general chairperson; Harold Berger, Industrial Quality, Inc., was technical chairperson; and Leonard Mordfin, NIST, served as arrangements chairperson.

Contents

Overview

Nondestructive testing (NDT), the examination of materials in ways that do not impair the intended uses of the materials, represents technology central to the concept of improved quality. Although the quality of materials, components, and products has always been important, it is clear that recent shifts in world trade and the growing awareness of the life-cycle costs of products has resulted in an increased appreciation of quality concepts and NDT. In order to achieve a better understanding of the role of NDT standards and their impact on world trade, ASTM Committee E-7 on Nondestructive Testing organized a three-day Symposium on Nondestructive Testing Standards II: New Opportunities for Increased World Trade Through Accepted Standards for NDT and Quality. The Symposium was held 9–11 April 1991 at the National Institute of Standards and Technology (NIST) in Gaithersburg, Maryland, under the joint sponsorship of Committee E-7 and NIST, and in cooperation with the American Society for Nondestructive Testing and the American Welding Society. James Borucki, chairman of Committee E-7, served as General Chairperson of the Symposium.

This special technical publication (STP) presents peer-reviewed versions of most of the papers presented at the Symposium. The title of the book was changed from the Symposium title to *Nondestructive Testing Standards—Present and Future* during the editing process when it became evident that the authors of the papers, almost without exception, had provided us with knowledgeable projections of what we may expect to see in the way of new NDT standards in the years ahead. We feel that this aspect of the book may well represent its most unique value, coming at a time when American industry is focused on quality considerations and European Community standards are becoming an additional factor in international trade.

The book has been divided into four sections: (1) NDT Standards: The ASTM Program; (2) NDT Standards: NIST, DoD, ASME, SAE, ISO, EC; (3) NDT Personnel Qualification: Here and Abroad; (4) NDT Standards: Advanced Applications.

NDT Standards: The ASTM Program

The standards development program of ASTM Committee E-7 is the largest and most comprehensive standards program for NDT in the country. It is appropriate, therefore, that the first section is devoted to this program.

The first paper, by Borucki, in addition to serving as a preamble to the symposium as a whole, provides an introduction to ASTM's NDT program. It describes the organization and operations of Committee E-7 and its excellent record of accomplishment in the development of NDT standards since its inception more than 50 years ago. Not content to discuss this sterling record, Borucki also discusses the ongoing efforts of the Committee to improve its operations in order to be even more responsive to industrial needs for NDT standards in the years ahead.

The next three papers address the Committee's work in radiologic NDT. Subcommittee E07.01 on Radiology (X and Gamma) Method evolved from an earlier committee on radiography that constituted one of the nuclei around which E-7 was originally established. As described by Graber, the longtime chairman of E07.01, this subcommittee has been extremely prolific, constituting the foremost authority and source for radiographic NDT standards in the country. But the Subcommittee continues to keep ahead of industry's needs, continually seeking to develop documents that will facilitate the adoption and utilization of new technologies such as radioscopy, computed tomography, and the special techniques needed for effective examination of composite materials.

In the next paper, Jones and Goldspiel describe the use of reference radiographs to assist in the interpretation and evaluation of radiographic images. They review the sets of reference radiographs which have been developed by and for Subcommittee E07.02 on Reference Radiographs and which are disseminated by the Society, and then go on to discuss the challenges which the future holds for this essential activity, e.g., composites, ceramics, and exotic alloys which exhibit different radiographic characteristics, and advanced radioscopic imaging systems in which radiologic interpretation is performed using video displays rather than traditional film viewers.

Committee E-7's program to develop standards for radiologic NDT is not limited to X- and gamma-ray radiology. In the next paper, Brenizer discusses the activities relating to neutron radiology. He reviews the basic physics of the method and describes the existing ASTM standards on the subject and the new standards under development, their benefits to industry, and the ongoing efforts to promulgate similar documents through the International Organization for Standardization (ISO).

Perhaps the most commonly used NDT methods are those used to detect surface flaws, namely, liquid penetrant testing and magnetic particle testing. ASTM's program to develop standards for these methods is described by Fenton, the chairman of Subcommittee E07.03 on Liquid Penetrant and Magnetic Particle Methods. He points out that the traditional output of the activity, i.e., standard guides, standard practices, and reference photographs, will soon be supplemented by *specifications* which the Subcommittee is now preparing. These new documents parallel respective military specifications and are expected to replace them in accordance with the Defense Department's policy to adopt nongovernment standards and specifications in lieu of military documents wherever possible.

The next paper traces the historic development of ultrasonic techniques for NDT. In it, Van Valkenburg, chairman of Subcommittee E07.06 on Ultrasonic Method during its most productive period, shows how the ultrasonic testing standards developed in this and other ASTM committees and elsewhere have paralleled the technology for more than four decades and continue to do so now, addressing topics such as the detection of intergranular stress-corrosion cracking and the computerized transfer of ultrasonic test data.

One of the newer NDT methods, acoustic emission testing, is discussed by Jolly in the next paper. He reviews the development of the technology and outlines its areas of application before describing the standards—including test methods, practices, and guides—that have been prepared by Subcommittee E07.04 on Acoustic Emission Method and other organizations. This discussion is followed by a summary of the new standards, presently under development, that address emerging topics such as acousto-ultrasound.

McEleney's paper on electromagnetic (eddy current) testing begins with an extensive summary of the basic principles and applications of this NDT method before proceeding into a discussion of the different types of standards that have been developed and the various product forms for which they are useful. As longtime chairman of Subcommittee E07.07 on Electromagnetic Method, McEleney also includes an interesting review of the Subcommittee's history, which leaves the reader with an enhanced appreciation of many of the Subcommittee's documents and how they were developed.

Subcommittee E07.10 on Other NDT Methods was established by McClung to serve as a home within Committee E-7 for standards activities on emerging or other NDT technologies that are not individually large enough to justify an independent subcommittee. He describes the program of the Subcommittee and the standards which it has developed, conveying the excitement that comes from harnessing—by means of standards—some of the newer NDT methods and related technologies.

Unlike the other technical subcommittees of Committee E-7, Subcommittee E07.09 on Nondestructive Testing Laboratories does not address a specific NDT method or methods. Rather, it is concerned with the administrative and operational requirements of a qualified NDT laboratory. In the next paper, Plumstead and Jaycox describe the standards which this subcommittee has developed and some of the rationales upon which they are based. They succeed in generating an appreciation for the value of such standards as well as for the enormous difficulties involved in their development.

The last paper in this section of the book was not part of the symposium upon which this STP is based. Rather, it was presented at an earlier ASTM symposium dealing with the standardization of technical terminology. However, the paper is included here because it addresses an important aspect of Committee E-7's standardization program for NDT, namely, the development of standard terminologies. In the paper, McKee describes the process which was used to combine separate glossaries, each dealing with a single NDT method, into a single, consistent, terminology compilation for all of Committee E-7's documents. The process was actually completed after the paper was prepared; the result is ASTM E 1316: Terminology for Nondestructive Evaluations.

NDT Standards: NIST, DoD, ASME, SAE, ISO, EC

The second section of this volume is devoted to NDT standards development activities outside of ASTM. The first paper, by Birnbaum, Eitzen, and Mordfin, surveys the NDT standards developed at NIST (formerly the National Bureau of Standards) since the previous ASTM Symposium on Nondestructive Testing Standards in 1976. NIST's standards are primarily measurement standards, as opposed to the documentary standards developed by ASTM and others, and this paper delves into the theoretical and experimental bases of the standards.

The Defense Department's standards development program for NDT is described in the next paper by Strauss. This program involves close collaborations between the Department, NIST, and various nongovernment standards-writing organizations, including ASTM and SAE. Strauss describes the process used to adopt nongovernment standards in lieu of military standards and the criteria which the nongovernment documents must meet in order for them to be acceptable to the DoD. The paper also addresses NDT standardization activities in JANNAF (the Joint Army-Navy-NASA-Air Force Interagency Propulsion Committee) and in some international military organizations such as NATO and ABCA.

The evolution of NDT requirements and standards in the ASME Boiler and Pressure Vessel Code is the subject of the next paper by Spanner. He describes the manner in which ASME Code rules are organized and the interrelationships between the several Code sections, the piping codes, and ASTM standards. Particular attention is focused on the significant changes in the Code relative to NDT that took place during the 1980s, and projections and trends for the 1990s.

The development of NDT standards for the aerospace industry is a particular function of Committee K in the Aerospace Materials Division of SAE International. As described by Cooper and Nethercutt, these documents, called Aerospace Materials Specifications, address various NDT methods and materials and are acquiring greater importance in light of the Defense Department's policy, cited above, of replacing military standards and specifications, where possible, with nongovernment consensus standards. The authors trace the history and the organization of the Committee and outline its plans to expand its efforts in order to better cover all of the major NDT methods.

The final two papers in this section address international standards for NDT. The paper by Mordfin briefly describes the process for developing standards which is used by ISO and the manner in which the United States participates in this process. The role of international standards in international trade and the importance of having international standards that are consistent with the practices of American industry are stressed, and a plea is made for greater support, from both government and industry, for enhanced U.S. participation in the development of international standards for NDT.

As part of the European Community's plan to become a single market by the end of 1992, an intensive effort is underway to harmonize the existing national regulations of the various countries in order to eliminate some barriers to trade. As described by Borloo, this effort is carried out under the European Committee for Standardization, in which one technical committee has been established specifically to focus on NDT, and another, on welding, which also addresses NDT issues. The paper reviews the organization and the program of work of these two technical committees. Borloo makes the point that, in the development of European standards for NDT, the technical committees consider documents of non-European origin (e.g., ASTM) as well as those from the European national standardizing bodies.

This section of the book concludes with a report, by Spanner, on a panel discussion that was held during the Symposium.

NDT Personnel Qualification: Here and Abroad

Because standards for NDT personnel qualification are different from standards for NDT methods, and because the subject of NDT personnel qualification and certification has generated enormous interest—and controversy—in recent years, a separate section of this book was assigned to this topic. Two papers are included. In the first, Wheeler reviews the status of NDT personnel qualification and certification from the point of view of various U.S. organizations and practitioners and compares this with the practices of some other countries. He also presents some interesting results of a survey to evaluate the spectrum of current U.S. attitudes on the subject and concludes that an international standard on NDT personnel qualification that is generally acceptable to the United States as well as to most other nations is now realistically achievable.

The ISO effort to develop such a standard is the subject of the next paper. In it, Zirnhelt describes the need for the standard and traces the historical background of the effort, including the various meetings and ballots which have served to narrow differences and approach consensus among the member countries. In keeping with his role as the chairman of the effort, he goes on to discuss, in a nonpartisan way, the influence of the European Community on the development of the standard and the national implications of such a document. The paper concludes with a review of some of the standard's more contentious issues and a glimpse at some of the related challenges that still lie ahead.

NDT Standards: Advanced Applications

The final section of the book deals with NDT standards for advanced applications. The first paper, by Berger and Hsieh, addresses the challenging need for standards to facilitate the transfer, exchange, and combination of data from computerized NDT equipment. Related documents are reviewed, and progress toward the development of the needed standards is described.

Most of today's NDT standards were developed with metals in mind, whether consciously or not. However, these standards will not be adequate for the high-temperature materials needed for tomorrow's automotive gas turbines and various aerospace applications. These applications and others like them will likely require structural ceramics and complex composite materials with stringent NDT needs that exceed current capabilities. Examples of these needs include the reliable detection of flaws less than 100 μm in size, and high-definition imaging systems. In an extremely thorough and well-documented paper, Vary elaborates on these and other needs and discusses the standards that ASTM must begin to address.

The next paper deals with an advanced application of NDT standards, rather than NDT standards for an advanced application. Based largely on his own observations, Plumstead describes the excessive and long-term costs to the construction industry that result from the all-too-common practice of using unqualified NDT subcontractors to perform the required nondestructive examinations. In response to this deficiency, he presents a thoughtful proposal for a system that would require prequalification of NDT subcontractors to ASTM E 543: Practice for Determining the Qualification of Nondestructive Testing Agencies, and other standards.

The last paper in the book is concerned with the problems involved in magnetic particle testing when inspection must simultaneously satisfy standards of different countries or organizations. As an approach to easing such difficulties, Stadthaus presents here a review of a new German guide which provides comparisons and comments on several important national and international standards for magnetic particle testing.

This volume, which provides a comprehensive review of NDT standards from many points of view, serves as a valuable update to the NDT standards review symposium held in 1976. That symposium, the proceedings of which were published in ASTM STP 624, represented a serious effort by the NDT community to examine its standards to see if there were topics that had not been satisfactorily addressed. One of the results was greater attention being paid, in the ensuing years, to requirements for quantitative NDT measurements. In the present symposium, the pervasive global thinking seemed always to move the focus from the current status of NDT standards to the needs of the future and the increasing role that NDT standards will play in world trade. It is the hope of the symposium organizing committee and the symposium sponsors that these papers will serve to provide useful guidance and direction to our continuing efforts to develop new and improved NDT standards and thereby enhance our capabilities for NDT measurements and the quality of our products.

Harold Berger

Industrial Quality, Inc.,
  Gaithersburg, MD; technical
  chairperson and editor

Leonard Mordfin

National Institute of Standards and Technology,
  Gaithersburg, MD; arrangements
  chairperson and editor

NDT Standards: The ASTM Program

James S. Borucki[1]

Overview of ASTM E-7 Nondestructive Testing Standards

REFERENCE: Borucki, J. S., **"Overview of ASTM E-7 Nondestructive Testing Standards,"** *Nondestructive Testing Standards—Present and Future, ASTM STP 1151,* H. Berger and L. Mordfin, Eds., American Society for Testing and Materials, Philadelphia, 1992, pp. 9–14.

ABSTRACT: The American Society for Testing and Materials Committee E-7 has developed and implemented nondestructive testing standards for the NDT industry for well over 50 years. Today there are more than 125 ASTM NDT standards available covering all major nondestructive testing disciplines with several more under development. The work of Committee E-7 continues in the 1990s as we strive to respond to the ever-growing needs of the NDT industry to establish new standards and improve on existing documents.

This publication focuses on the realization that a myriad of NDT standards exist, that they originate in several organizations worldwide, and that there is a serious need for improved global standardization. It is a goal of Committee E-7 to gain a better understanding of worldwide NDT standards needs and to establish better liaison and cooperation with these other organizations so that we can better respond to the needs.

KEY WORDS: standards, proposal, emergency standards, test method, nondestructive testing

The American Society for Testing and Materials has been involved in the development of nondestructive testing (NDT) standards for many years. As a matter of fact, ASTM was the first industrial society in the United States to recognize radiography as a method of routine testing. It was included as a part of X-ray metallography in Subcommittee VI of Committee E-4 on Metallography. This subcommittee originated in 1925. In 1938, because of the growing commercial importance of radiographic inspection, ASTM authorized the formation of a new standing Committee E-7.

Committee E-7 has today a working membership of over 325 members. From its inception over 50 years ago, Committee E-7 has grown from a method standards committee on radiography to its present day operation that covers such important NDT technologies as radiology, acoustic emission, electromagnetic, leak detection, liquid penetrant, magnetic particle, ultrasonics, thermoelectrics, and thermographic and other emerging NDT methods.

Today there are over 125 standards on NDT under ASTM jurisdiction. In this context, the term "standard" refers to documents including test methods, definitions, standard practices, classifications, and specifications.

ASTM NDT Standards Development Program

The American Society for Testing and Materials program for developing nondestructive testing standards resides primarily in Committee E-7 on Nondestructive Testing. Some prod-

---

[1] Vice president and director of Ardrox, Inc., North American NDT Systems Business Group, 16961 Knott Ave., La Mirada, CA 90638.

uct committees do include certain NDT practices in their product standards, but the main focus on NDT standards development rests within Committee E-7.

The many activities necessary in the development of standards documents include industry surveys to determine interest, needs, and practices; performance of research and development studies through extensive cooperative studies in both government and private organizations, document preparation, and round robin testing.

Committee E-7 is always responsive to requests from other standards committees and organizations such as the American Society of Mechanical Engineers (ASME), International Organization for Standardization (IOS), American National Standards Institute (ANSI), American Society of Nondestructive Testing (ASNT), Department of Defense (DOD), and ASTM product committees, as the need arises.

The general scope of Committee E-7 activities includes the promotion of knowledge, the advancement of NDT technology, and the stimulation of research in the nondestructive testing of materials and the extension of NDT methods to the solution of engineering problems. Also included is the interpretation and classification of the results of such nondestructive testing, but without prejudice to the jurisdiction of appropriate ASTM product committees over their respective products. E-7 efforts also include the coordination and review of NDT documents initiated by other committees as well as the maintenance of appropriate liaison (including coordination and/or consultation where desirable) with other ASTM technical committees and outside organizations concerned with nondestructive testing.

Over the years many benefits to industry have resulted from these ASTM NDT consensus standards. Needless to say, we must continue to strive to improve and expand our standards as well as provide for more relevant, quantitative, and reproducible results.

Participation in the various E-7 subcommittees, sections, and task groups is entirely voluntary, and broad-based industry-wide involvement is always encouraged. Traditionally, E-7 nondestructive testing documents have represented true consensus documents. Consensus is enhanced by joint participation of producers, consumers, and other interested parties.

ASTM Committee E-7 Structure

Committee E-7 is organized into appropriate administrative and technical subcommittees representing the various nondestructive testing disciplines as indicated in Tables 1 and 2.

Administrative Subcommittees

The administrative subcommittees of Committee E-7 act in support of the overall committee and its various technical subcommittees. Although it is not the function of these administrative groups to write standards, they do serve the vital role of organizational administration.

TABLE 1—*Administrative subcomittees of E7.*

E07.90—Executive
E07.91—USA ISO\TC135 on NDT
E07.92—Editorial Review
E07.93—Illustration Monitoring
E07.95—Long Range Planning
E07.96—Awards
E07.98—New Methods Review
E07.99—Liaison

TABLE 2—*Technical subcommittees of E07.*

E07.01—Radiology (X and Gamma) Method
E07.02—Reference Radiographs
E07.03—Liquid Penetrant and Magnetic Particle Methods
E07.04—Acoustic Emission Method
E07.05—Radiology (Neutron) Method
E07.06—Ultrasonic Method
E07.07—Electromagnetic Method
E07.08—Leak Testing Method
E07.09—Materials Inspection and Testing Laboratories
E07.10—Emerging NDT Methods

Subcommittee E07.90 is the Executive Council which serves as the steering committee for general business matters such as establishing new work scopes, establishing new subcommittees, approving general membership, approval of subcommittee chairmen appointments, approval of meeting sites, planning symposiums, and other ad hoc business matters.

Subcommittee E07.91 is the U.S. Technical Advisory Group for the International Organization for Standardization/Technical Committee 135 on NDT. This subcommittee coordinates and provides the technical expertise and participation of the United States in the International Organization for Standardization.

Subcommittee E07.92 on Editorial Review provides editorial support specific to the specialties of the E-7 technical subcommittees.

Subcommittee E07.93 on Illustration Monitoring works in conjunction with ASTM headquarters in reviewing the production of reference radiographs and other illustrations that are integral parts of E-7 standards.

Subcommittee E07.95 on Long Range Planning provides for constant monitoring of the efficiency of operation, seeks ways of resolving deficiencies of the various technical subcommittees, and establishes strategic planning.

Subcommittee E07.96 on Awards coordinates and implements the various subcommittees' nominees for E-7 and ASTM member recognition awards. In addition, the Awards Committee coordinates all activities relative to Society and E-7 social functions.

Subcommittee E07.98 on New Methods Review maintains an active awareness of nondestructive testing methods not covered currently by standardization activity and makes recommendations when standards in additional methods appear to be necessary.

Subcommittee E07.99 on Liaison coordinates all liaison activities with other technical organizations such as ASME (American Society of Mechanical Engineers), SAE/AMS (Society of Automotive Engineers/Aerospace Material Specifications), AWS (American Welding Society), ASNT (American Society for Nondestructive Testing), as well as other ASTM Committees requiring NDT expertise.

Technical Subcommittees

The various technical subcommittees of Committee E-7 originate, revise, and serve as the overall custodian for the various NDT documents under their respective jurisdictions.

E07.01 on the Radiology (X and Gamma) Method is concerned with the formulation and standardization of practices, methods, and guides used for industrial X-ray and gamma ray radiology. These documents address methodology, terminology, test techniques, materials, and product forms for which X-ray and gamma ray radiology is used as the test method.

E07.02 on Reference Radiographs is concerned with the development of reference radiograph documents applicable to the evaluation of metals and related materials in various stages and forms of fabrication. Also included are reference radiographs for semiconductors and associated electronic devices.

E07.03 on the Liquid Penetrant and Magnetic Particle Methods is concerned with the formulation and standardization of test methods, terminology, practices and guides, and the promotion of knowledge relating to liquid penetrant and magnetic particle testing.

E07.04 on the Acoustic Emission Method is concerned with the formulation and standardization of test methods, practices, guides, and terminology; and the promotion of knowledge relating to acoustic emission examination.

E07.05 on the Radiology (Neutron) Method is concerned with the formulation and standardization of test methods, practices, guides, quality indicators, and terminology relating to neutron radiology and the promotion of knowledge in this nondestructive testing method.

E07.06 on the Ultrasonic Method is concerned with the formulation and standardization of test methods, terminology, practices, and guides; and the promotion of knowledge relating to ultrasonic testing applied to detection of discontinuities, measurement of dimensions, and characterization of the properties of materials.

E07.07 on the Electromagnetic Method is concerned with the formulation and standardization of test methods, practices, guides, and terminology, and the promotion of knowledge relating to electromagnetic (eddy current) testing. Activities in the magnetic method of coating thickness measurement, the fringe flux method, and d-c and a-c potential (electric current) method are included.

E07.08 on the Leak Testing Method is concerned with the formulation and standardization of test methods, practices, guides and terminology, and the promotion of knowledge relating to leak measurement, location, and monitoring.

E07.09 on Materials Inspection and Testing Laboratories is concerned with standards that will guide the organization and operation of nondestructive examination and testing agencies, to develop practices, including minimum requirements regarding the facilities and equipment of nondestructive examination and testing agencies, to list pertinent NDT methods for the materials to be examined, and to work with other technical committees of ASTM and other organizations having an interest in this field toward the continual upgrading of the technical and ethical standards for nondestructive testing agencies/laboratories.

E07.10 on Emerging NDT Methods is concerned with the formulation and standardization of nondestructive testing by methods for which no other subcommittee exists within Committee E-7. Typical examples in the initial organization of the Subcommittee were activities for optical holography, thermoelectric metal sorting, and optical/visual NDT. As such, it serves as an umbrella organization for emerging technologies for which the level of necessary activity does not justify an independent subcommittee.

Addressing the Need for NDT Standards

As discussed earlier, Committee E-7 on Nondestructive Testing is well positioned to address the various standards needs of the NDT industry. Its technical subcommittees, sections, and task groups are capable of executing standards projects within their areas of jurisdiction.

It would be difficult, at best, for an organization such as Committee E-7 to operate passively, that is, to "sit and wait" for the "phone to ring." Fortunately, the participating members of E-7 are active technical experts representing a variety of industrial, governmental, and institutional interests. These experts work very diligently in our various E-7 technical subcommittees assuring that existing documents are kept current with changing technology and changing inspection practices. In addition, new documents are constantly initiated in response to industry requests.

Recognition of the need for a new or improved standard can come from a variety of sources. The most frequent source, however, is usually the members themselves, who are the technical experts within the subcommittee structure who recognize both the need and the mechanism for satisfying the need. Other major sources alerting a need for specific standards include other technical organizations, societies, government agencies, private industry and, of course, other ASTM technical committees.

The decision and timing for initiating a standard document project is entirely under the jurisdiction of the respective technical subcommittee as long as the rules of standards development are followed relative to the E-7 scope.

Obviously, not every request of E-7 represents a genuine need for an industry-wide consensus standard. Requests, therefore, are screened and evaluated to determine if the need really exists for the preparation of a new standard or to determine if the need can be fulfilled by existing documents or by revision of an existing document to include the new requirement. It is also important to screen and evaluate requests to assure that proprietary commercial interests are not self serving.

Mechanism for Preparing and Approving ASTM NDT Standards

As stated earlier, requests to establish new standards or to reissue/improve existing standards may come from a variety of sources. Whatever the source, however, each must be fully justified.

The mechanism for screening, preparation, and approval hasn't really changed much over the past 50 years. What has changed is the growth of the NDT industry and the development of a myriad of new NDT method standards. After recognizing the need, the Executive Council of Committee E-7 (E07.90) will direct requests for a specific NDT standard to the specific technical subcommittee which represents the technical expertise for that particular NDT methodology. This technical subcommittee will then review the request, assess the technical requirements for the document, and proceed accordingly with its development if sufficient documentation and technical justification are available. There are many instances where additional R & D and round robin testing is required in order to develop the documentation necessary to support the establishment of the new standard document. The actual work of preparing the first draft of a given document is usually done by a task group of technical experts familiar with the technology.

After a new draft standard has been prepared, it is ready for subcommittee balloting. Subcommittee members must exercise their voting choices, i.e., affirmative, negative, abstention. Ballot results are then tabulated and evaluated to assure that the 60% minimum vote response is attained.

All comments are reviewed by the drafting task group and incorporated into the document if considered valid. Likewise, negative votes are reviewed and a determination made as to whether or not they are regarded as persuasive or nonpersuasive. Valid recommended technical changes require reballoting.

Following a successful subcommittee ballot, the document is ready for the main committee ballot where the full E-7 member body has the opportunity to review the draft and cast its votes. The voting requirements are the same as those cited for the subcommittee ballot process. Again all negatives must be resolved and valid comments incorporated into the document. The final step in the approval process is the ASTM society-level ballot process.

The NDT standards development process described above and its mechanism for development and approval clearly demonstrate the ASTM voluntary consensus process. This process has worked successfully for over 50 years and continues to be vital to the nondestructive testing industry worldwide.

Fulfilling Global Market NDT Standards Needs

The world demand for high-quality products bodes well for those countries and companies whose products make use of accepted standards for nondestructive testing. One of the major problems we face, however, during this period of "global standardization" is the determination of which NDT standards/specifications to adopt as international working standards. There is no question that there are a myriad of NDT method specifications emanating from several global sources. The problem we face, however, is how to improve on the quality of these documents, eliminate redundancies, and standardize where we can so as to be in harmony on a corporate, national, and global basis. Significant progress is being made in this regard through the participating efforts of ISO/TC 135, other technical organizations, government agencies, and private industry.

ASTM Committee E-7 Strives to Improve

Over the years, Committee E-7 has contributed greatly to industry in its response and development of NDT standards. Its standards development efforts and research and development programs have contributed greatly to the high level of sophistication in the science of nondestructive testing. There are, however, many areas where improvements could be made to improve the overall quality and efficiency of our operation.

To this end we are attempting to address key areas for further improvement which can impact the usefulness of our ASTM NDT standards:

1. Assure that the best current technology is known, understood, and applied.
2. Assure that the technical facts in the document are accurate.
3. Assure prompt response to requests for assistance or in recognizing the need for a new standard.
4. Assure active liaison with government and industry.
5. Assure active participation with other standards writing bodies.
6. Shorten the time interval between initiation and the publication of standards.

Summary

From papers in this symposium you will learn of the many activities and growth areas that ASTM Committee E-7 has to offer. You will learn of the NDT standards programs of the National Institute of Standards and Technology, the International Organization for Standardization, the U.S. Department of Defense, the American Society of Mechanical Engineers, the Society of Automotive Engineers Aerospace Material Specifications, National Aeronautics and Space Administration, Federal Aviation Administration, and the international industrial community—each expressing its respective position on NDT standards needs.

This symposium truly represents a unique opportunity to improve our nondestructive testing international standards development process.

Harold C. Graber[1]

Present and Future ASTM Standards Developments in the Radiology (X and Gamma) NDE Test Method

REFERENCE: Graber, H. C., **"Present and Future ASTM Standards Developments in the Radiology (X and Gamma) NDE Test Method,"** *Nondestructive Testing Standards—Present and Future, ASTM STP 1151,* H. Berger and L. Mordfin, Eds., American Society for Testing and Materials, Philadelphia, 1992, pp. 15–22.

ABSTRACT: ASTM Subcommittee E07.01 on Radiology has been focusing efforts in expanding the development of standards in several different areas. In the past several years, standards addressing the radiologic examination of product forms, radioscopic imaging, and radiologic system variables have been published. An overview of these standards will be presented.

Standards work identified for future development include, for example, performance standards and product form examination for radioscopy, radiologic examination of composite materials and tutorial, and performance standards for computed tomography. An overview of the current status and committee activity for future work will be presented.

KEY WORDS: radiology, radiography, radioscopy, radiologic, nondestructive examination, X-ray, gamma ray

The ASTM Subcommittee E07.01 on Radiology has historical significance related to the overall progress Committee E07 on Nondestructive Testing has made since its inception in 1936.

The first report of the E07 Committee states in part:

The American Society for Testing and Materials was the first industrial Society in this country to recognize radiography as a method of routine testing. It was included as a part of X-ray metallography in Subcommittee VI of Committee E-4 on Metallography, which was organized in 1925. In 1937, because of the growing importance of this method, the Society authorized the formation of a new Standing Committee devoted to radiography. Preliminary organization was effected in the fall of that year. Permanent officers were elected in February, 1938, and the first regular meeting of the Committee was held in Rochester, New York, in March, 1938.

Since the beginning, approximately 54 years ago, Committee E07 and Subcommittee E07.01 have been blessed with the privilege of having notable experts in the field of radiography serve on the E07 Committee as committee or subcommittee chairmen. To name a few, H. H. Lester, Jim Bly, Alexander Gobus, Ralph Turner, Dan Polansky, Sol Goldspiel, Ed Criscuolo, Harold Berger, and a host of others too numerous to mention. To these individuals, we owe a great deal of credit for their individual contributions and dedication, which resulted in the progress and advancement of ASTM standards on radiology.

[1] Manager, Central Inspection and Testing, Babcock Wilcox Co., Barberton, OH 44203.

Present Standards

As mentioned in the Introduction, in the early beginning in the mid-1930s, the Committee focused on radiographic testing of metal castings. One of the first standards published was Tentative Methods of Radiographic Testing of Metal Castings, ASTM Designation E 15-39T, issued 1939.

From this example of a standard issued 52 years ago, the state of the art has grown and branched out to many facets. In fact, today the terminology has changed from "radiography" to "radiology." This terminology change was necessitated by the expansion of the basic test method techniques which provide a means of obtaining or recording information other than on a film. Figure 1 provides a generic family tree of terminology which illustrates the expansion of the technology.

In order to provide an insight and understanding of present Subcommittee E07.01 activity, Fig. 2 provides the current organization structure.

The following is a descriptionof the published standards in the *ASTM 1990 Annual Book of Standards,* Section 3, Volume 03.03, which are the responsibility of the respective sections.

GENERIC FAMILY TREE
OF TERMINOLOGY

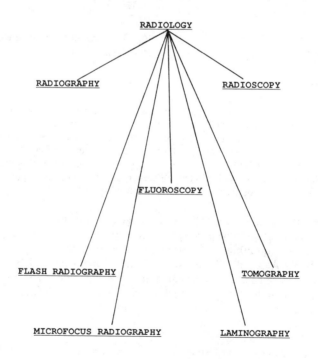

Fig. 1.

ASTM SUBCOMMITTEE E-7.01
ON RADIOLOGY

CHAIRMAN - H. C. GRABER
SECRETARY - J. J. MUNRO, III

THE SUBCOMMITTEE CONSISTS OF FIVE
SECTIONS:

- E-7.01.01   RADIOGRAPHY
  CHAIRMAN: T. L. BAILEY

- E-7.01.02   RADIOSCOPY
  CHAIRMAN: G. B. NIGHTINGALE

- E-7.01.03   COMPUTED TOMOGRAPHY
  CHAIRMAN: J. STANLEY

- E-7.01.04   TERMINOLOGY/GLOSSARY
  ACTING CHAIRMAN: H. C. GRABER

- E-7.01.05   APPLICATIONS
  CHAIRMAN: J. K. AMAN

Fig. 2.

Section E07.01.01 Radiography

Methods

E 1114-86   **Method for Determining the Focal Size of Iridium-192 Industrial Radio-graphic Sources**
This method covers the determination of the focal size of an Iridium-192 radiographic source. The determination is based upon measurement of the image of the iridium metal source in a projection radiograph of the source assembly and comparison with the measurement of the image of a reference sample in the same radiograph.

E 1165-87   **Test Method for Measurement of Focal Spots of Industrial X-Ray Tubes by Pinhole Imaging**
This test method provides instructions for determining the length and width dimensions of line focal spots in industrial X-ray tubes. This determination is based on the measurement of an image of a focal spot that has been radio-graphically recorded with a "pinhole" projection/imaging technique.

E 746-87   **Method for Determining Relative Image Quality Response of Industrial Radiographic Film**
This test method covers the determination of the relative image quality response of industrial radiographic film when exposed to 200-kV X-rays. The evaluation of the film is based upon the threshold visibility of penetrameter holes in a special image quality indicator (IQI). Results for a given film type may vary, depending upon the particular development system used. It is, therefore, necessary to state the development system and geometric conditions used in this determination. By holding the technique parameters (except exposure time) and processing parameters constant, the image quality response of radiographic film may be evaluated on a relative basis.

E 747-90   **Test Method for Controlling Quality of Radiographic Testing Using Wire Penetrameters**
This test method covers the radiographic examination of materials for discontinuities using wire penetrameters as the controlling image quality indicator for the material thickness range from 6.4 to 152 mm (0.25 to 6.0 in.). Requirements expressed in this method are intended to control the reliability or quality of radiographic images and are not intended for controlling acceptability or quality of materials or products.

E 1025-89   **Practice for Hole-Type Image Quality Indicators Used for Radiography**
This practice covers the design and material grouping classification of hole-type image quality indicators used to indicate the quality of radiographic images.

E 801-90   **Practice for Controlling Quality of Radiographic Testing of Electronic Devices**
This practice relates to the radiographic examination of electronic devices for internal discontinuities and extraneous material within cavities or encapsulating materials. Requirements expressed in this practice are intended to control the reliability or quality of the radiographic images and are not intended for controlling the acceptability or quality of the electronic devices radiographed.

Practices
E 1079-85(91)   **Practice for the Calibration of Transmission Densitometers**
This practice covers the calibration of transmission densitometers used to perform radiographic film density measurements.

Guides
E 94-89   **Guide for Radiographic Testing**
This guide covers satisfactory X-ray and gamma-ray radiographic testing as applied to industrial radiographic film recording. It includes statements about preferred practice without discussing the technical background which justifies the preference. A bibliography of several textbooks and standard documents of other societies is included for additional information on the subject.

E 999-90   **Guide for Controlling the Quality of Industrial Radiographic Film Processing**
This guide establishes guidelines that may be used for the control and maintenance of industrial radiographic film processing equipment and materials. Effective use of these guidelines aids in controlling the consistency and quality of industrial radiographic film processing.

E 1254-88   **Guide for the Storage of Radiographs and Unexposed Industrial Radiographic Films**
This guide may be used for the control and maintenance of industrial radiographs and unexposed films used for industrial radiography.

E 1390-90   **Guide for Illuminators Used for Viewing Industrial Radiographs**
This guide provides the recommended minimum requirements for illuminators used for viewing industrial film radiographs using transmitted light.

Section E07.01.02 Radioscopy

Practices
E 1255-88   **Practice for Radioscopic Real-Time Inspection**
This practice provides application details for radioscopic examination using penetrating radiation. This includes real-time radioscopy and, for the purposes of this standard, radioscopy where the motion of the test object must be limited (commonly referred to as near-real-time radioscopy). Since the techniques involved and the applications for radioscopic examination are diverse, this practice is not intended to be limiting or restrictive, but rather to address the general applications of the technology and thereby facilitate its use.

E 1411-91   **Practice for the Qualification of Radioscopic Systems**
This practice provides test and measurement details for measuring the performance of X-ray and gamma-ray radioscopic systems. Radioscopic examination applications are diverse; therefore, radioscopic system configurations are also diverse and constantly changing as the technology advances.

Guides
E 1000-89   **Guide for Radioscopic Real-Time Imaging**
This guide is for tutorial purposes only and to outline the general principles of radioscopic imaging. It describes practices and image quality measuring systems for real-time and near-real-time nonfilm detection, display, and recording of radioscopic images. These images, used in materials inspection, are generated by penetrating radiation passing through the subject material and producing an image on the detecting medium. Although the described radiation sources are specifically X-ray and gamma ray, the general concepts can be used for other radiation sources, such as neutrons. The image detection and display techniques are nonfilm, but the use of photographic film as a means for permanent recording of the image is not precluded.

Section E07.01.03 Computed Tomography

No standards published to date.

Section E07.01.04 Terminology/Glossary

E 586-90a   **Standard Terminology Relating to Industrial Radiology**

E 1316-90   **Terminology for Nondestructive Evaluations**
This standard defines the terminology used in the standards prepared by the E07 Committee on Nondestructive Testing. These nondestructive testing (NDT) methods include: acoustic emission, electromagnetic testing, gamma and X-radi-

ography, leak testing, liquid penetrant testing, magnetic particle testing, neutron radiography and gaging, ultrasonic testing, and other technical methods. Section A defines terms that are common to all NDT methods, and the subsequent sections include the terms pertaining to a specific NDT method.

Section E07.01.05 Applications

Methods

E 1032-85   Method for Radiographic Examination of Weldments
This method provides a uniform procedure for radiographic examination of weldments using industrial radiographic film. Requirements expressed in this method are intended to control the quality of the radiographic process and are not intended for controlling acceptability or quality of welds.

E 1030-90   Test Method for Radiographic Examination of Metallic Castings
This test method provides a uniform procedure for radiographic examination of metallic castings that will produce satisfactory and consistent results upon which acceptance standards may be based. This test method covers the radiographic examination of materials for discontinuities with the use of radiographic film as the recording medium. Requirements expressed in this test method are intended to control the quality of the radiographic process and are not intended for controlling the acceptability or quality of materials or products.

E 1161-87   Test Method for Radiographic Testing of Semiconductors and Electronic Components
This test method provides a standard procedure for nondestructive radiographic examination of semiconductor devices, electronic components, and the materials used for construction of these items. This test method covers the radiographic testing of these items for possible defective conditions such as extraneous material within the sealed case, improper internal connections, voids in materials used for element mounting, or the sealing glass, or physical damage.

E 1416-91   Test Method for Radioscopic Examination of Weldments
This test method provides a uniform procedure for radioscopic examination of weldments. The method is used for the detection of discontinuities. The method facilitates the examination of a weld from several directions. The radioscopic techniques described in the method provide adequate assurance for defect detectability; however, for special applications, special techniques may be required.

The above listing totals 20 published standards which have evolved since the early beginning in 1936. The number may not seem to be very impressive; however, one must realize the vast amount of voluntary work involved in most cases to conduct round robin tests to obtain empirical data, validate the data, and process through the balloting procedure to ensure that the high quality of ASTM standards is achieved.

Future Direction and Activity

Each of the sections in Subcommittee E07.01 are actively pursuing and addressing customer needs for the future. Following is a brief description of the current section activities related to the revision of published standards and/or development of new standards for the future.

Section E07.01.01 Radiography

Revisions to Existing Standards

Standard E 1025   *Revision*—Addressing the inclusion of requirements for utilizing thinner plaques, e.g., several mils in thickness, for radiographing thin sections.

Standard E 94   *Revision*—Addressing revisions for energy selection and radiographic equivalence factors for several metals.

Proposed New Standards

Image Quality Indicators for Composite Materials.

Standard for Determining Relative Image Quality Response of Industrial Radiographic Film at Low Energies, e.g., below 200 kV. This is a companion standard to E 746.

Standard for Use of Unsharpness Indicators—Considering use of the British Standard BS-3971 indicator.

Standard for Film Process Classification of Industrial Radiographic Film.

Standard for Microfilming Industrial Radiographs.

Standard for Determining Relative Image Quality Response of Industrial Radiographic Film at High Energies, i.e., above 1 MV.

Standard for Determining Focal Spot Size of High Energy X-Ray Units.

Standard for Determining Focal Spot Size of Microfocus X-Ray Units.

Standard for Qualification of a Radiographic System.

Standard for Energy Calibration in the range of 50–300 kV.

Section E07.01.02 Radioscopy

Proposed New Standards

Standard for Analog Image Storage and Retrieval.

Standard for Storage Media Used to Store Radioscopic Images.

Standard for Data Fields for Computerized Transfer of Digital Radiologic Test Data.

Standard for IQIs Used for Radioscopic Examination.

Section E07.01.03 Computed Tomography

Proposed New Standards

Standard Guide for Computed Tomography Imaging.

Section E07.01.04 Terminology/Glossary

Incorporation of Computed Tomography Terms (approximately 100) in E 1316.

Section E07.01.05 Applications

Proposed New Standards

Standard for Flaw Sizing Determination from the Resultant Image, e.g., radiography or radioscopy.

Standard for Radiography of Composites.

Standard for Radioscopy of Castings.

Standard for Radioscopy of Electronic Components.

Standard Practice for Computed Tomography Examination.

All of the above listed proposed standards are in various stages of the developmental process. This listing demonstrates the emerging technology prevalent today in the field of radioscopic examination.

Summary

From the author's perspective, the information provided in this paper illustrates the significance of the advancements in the state of the art and also the new and emerging technologies, e.g., radioscopy and computed tomography. Further, an assessment of the tasks addressed by the Subcommittee illustrates the need for continued improvement and expansion of standards for radiographic examination methodology, from which the Committee originated.

Subcommittee E07.01 on Radiology is dedicated to serving Committee E07's mission, whose primary purpose is to develop and promulgate voluntary consensus standards related to the methodology of NDT. Further, the Subcommittee is dedicated to maintaining E07 Committee's position as the foremost authority and source for NDT methodology standards in the USA.

Acknowledgment

The author would like to acknowledge the members of Subcommittee E07.01 on Radiology for their effort, dedication, and contributions. Their voluntary effort, dedication, and contributions are reflected by the activity and accomplishments cited herein.

Thomas S. Jones[1] and Solomon Goldspiel[2]

Reference Radiograph Standards: Past, Present, and Future

REFERENCE: Jones, T. S. and Goldspiel, S., **"Reference Radiograph Standards: Past, Present, and Future,"** *Nondestructive Testing Standards—Present and Future, ASTM STP 1151,* H. Berger and L. Mordfin, Eds., American Society for Testing and Materials, Philadelphia, 1992, pp. 23–33.

ABSTRACT: Radiographic interpretation is a complex art involving visual discriminations of subtle radiographic details. The unambiguous communication of these details between purchaser and supplier has frequently proven a difficult task. Reference radiographs supply a direct visual reference to a range of discontinuity types and severity levels, providing a visually communicated basis for evaluating components to an agreed upon acceptance level. The first set of widely used reference radiographs were assembled by the U.S. Bureau of Engineering in 1938. This set of production radiographs illustrated a variety of defect types in steel castings and was adopted by ASTM in 1952 following the formation of Subcommittee E7.02 on Reference Radiographs. ASTM currently maintains 13 reference radiograph documents covering castings of a wide variety of materials and steel weldments. All current documents are produced using made-to-purpose hardware containing intentionally produced discontinuities of various types and severities. The future holds a great number of challenges for this activity due primarily to two factors. First is the rapid development of many new materials for aerospace and other applications. These materials include composites, ceramics, and exotic alloys. Second is the development and use of advanced radioscopic imaging systems in which radiologic interpretation is performed using video displays rather than the traditional film viewers. The development of reference radiological images appropriate to these systems offers many new challenges.

KEY WORDS: reference radiographs, castings, weldments, nondestructive testing, radiography, radiology, radioscopy

Why Reference Radiographs?

Industrial radiographs are direct graphical representations of the inspected components and, as such, often give the first impression of being easy to interpret. Nothing could be further from the truth. The industrial radiographer must make subtle discriminations between normal conditions, film artifacts, and component flaws. Flaw indications must further be evaluated against established acceptance criteria to determine whether they represent defect conditions. Radiographic indications can be very subtle and complex, involving networks of fine patterns. These patterns can be extremely difficult to describe in an unambiguous manner so that similar indications will be interpreted in the same way by multiple interpreters. Not only must the nature of the condition be accurately communicated, but the critical severity level must also be defined. Thus, radiographic interpretation is a learned skill involving careful discriminations of subtle radiographic details and evaluation of these details against an established acceptance level. The clear communication of these details between purchaser and supplier or

---

[1] Manager of research and engineering, Industrial Quality, Inc., 19634 Club House Rd., Suite 320, Gaithersburg, MD 20879.

[2] Consultant, 732 Gerald Ct., Brooklyn, NY 11235.

between engineering and quality control has frequently proven a difficult task. It is for these reasons that reference radiographs were developed. Reference radiographs supply a direct visual reference to a range of discontinuity types and severity levels, providing a visually communicated basis for evaluating components to an agreed upon acceptance level.

Historical Development of Reference Radiographs

During the late 1920s and early 1930s the use of X and gamma ray radiography was investigated for the examination of structural hardware, particularly castings and weldments in pressure vessel applications, in an effort to extend the service load capability for these structures [1,2]. In 1930, the U.S. Navy began requiring the radiographic examination of welded longitudinal and circumferential joints in boiler drums [3], and in 1931 the ASME Boiler Code made X-ray examinations mandatory for welded seams in power boiler drums and other pressure vessels designed for severe service conditions [2]. By 1933, Navy Department Specification 49Sl (Int) called for the radiographic examination of certain hull and pressure castings [1]. In these early applications the radiographic inspections were used to detect gross defects and to aid in the development of the casting processes [3]. As the use of radiographic inspection became more widespread, the need to standardize the interpretation and acceptance practices for particular types of components became clear.

During 1938, the Navy Department's Bureau of Engineering issued a set of reference radiographs titled "Gamma Ray Radiographic Standards for Steel Castings for Steam Pressure Service." This standard included radiographs showing acceptable, borderline, and unacceptable gradations of several types of discontinuities found in steel castings. Some of the unacceptable conditions were identified as being suitable for weld repair. Others indicated that no repairs would be permitted [3]. This document quickly became widely used by commercial organizations as well as by the Navy.

In 1940, the Navy began work on assembling a set of radiographs which covered the use of both X-ray and gamma ray techniques. These reference radiographs were adopted by the U.S. Bureau of Ships in 1942 and consisted of 31 plates, each having a gamma ray and X-ray radiograph of the same defect and represented both pressure and hull castings [3]. This document provided recognition that the radiation source type and energy can significantly affect the resulting image. This document also introduced the concept of graded discontinuities and provided a table indicating the use of the graded series in the evaluation of castings of varying criticality. Table 1 illustrates the designation of acceptable, borderline, and unacceptable defects for each class of casting.

Not all of the early efforts to standardize radiographic interpretation centered around the development of reference radiographs. In 1942, R. S. Busk proposed a method of grading porosity or microshrinkage in magnesium castings using the contrast between the unaffected and porous areas [4]. This approach used varied illuminator brightness and threshold visibility to grade the severity of the porosity.

Around 1944, the Army Air Corps Material Division in Dayton, Ohio released Air Corps Technical Report No. 4796 entitled "Aircraft Quality Casting Standards." This report included a number of films illustrating defects in castings and indicated the seriousness of each type of discontinuity. In some cases the report indicated that the seriousness of the condition would depend on the use of the component and that acceptability should be based on static tests of similar components [5].

In July, 1945, Lockheed Aircraft Corporation released LAC Specification 561, which illustrated acceptance levels for a variety of conditions. Acceptance levels for noncritical parts were chosen to produce a rejection rate of approximately 5%, which was considered tolerable. Acceptance levels for critical parts were based on statistics from production radiographs, phys-

TABLE 1—*Radiographic acceptance standards.*

Type of Defect	Plate	Class 1	Class 2	Class 3	Class 4	Class 5
Gas and	A1	Borderline	Acceptable	Acceptable	Acceptable	Acceptable
blowholes	A2	Unacceptable	Borderline	Acceptable	Acceptable	Acceptable
	A3	Unacceptable	Unacceptable	Borderline	Acceptable	Acceptable
	A4	Unacceptable	Unacceptable	Unacceptable	Borderline	Acceptable
	A5	Unacceptable	Unacceptable	Unacceptable	Unacceptable	Borderline
	A6	Unacceptable	Unacceptable	Unacceptable	Unacceptable	Unacceptable
Sand spots	B1	Borderline	Acceptable	Acceptable	Acceptable	Acceptable
and	B2	Unacceptable	Borderline	Acceptable	Acceptable	Acceptable
inclusions	B3	Unacceptable	Unacceptable	Borderline	Acceptable	Acceptable
	B4	Unacceptable	Unacceptable	Unacceptable	Borderline	Acceptable
	B5	Unacceptable	Unacceptable	Unacceptable	Unacceptable	Borderline
	B6	Unacceptable	Unacceptable	Unacceptable	Unacceptable	Unacceptable
Internal	C1	Borderline	Acceptable	Acceptable	Acceptable	Acceptable
shrinkage	C2	Unacceptable	Borderline	Acceptable	Acceptable	Acceptable
	C3	Unacceptable	Unacceptable	Borderline	Acceptable	Acceptable
	C4	Unacceptable	Unacceptable	Unacceptable	Borderline	Acceptable
	C5	Unacceptable	Unacceptable	Unacceptable	Unacceptable	Borderline
	C6	Unacceptable	Unacceptable	Unacceptable	Unacceptable	Unacceptable
Hot tears	D1	Unacceptable	Unacceptable	Unacceptable	Borderline[a]	Acceptable
	D2	Unacceptable	Unacceptable	Unacceptable	Unacceptable	Borderline[a]
	D3	Unacceptable	Unacceptable	Unacceptable	Unacceptable	Unacceptable
Cracks	E1	Unacceptable	Unacceptable	Unacceptable	Borderline[a]	Acceptable
	E2	Unacceptable	Unacceptable	Unacceptable	Unacceptable	Borderline[a]
	E3	Unacceptable	Unacceptable	Unacceptable	Unacceptable	Unacceptable
Unfused	F1	Unacceptable	Borderline	Acceptable	Acceptable	Acceptable
chaplets	F2	Unacceptable	Unacceptable	Borderline	Acceptable	Acceptable
	F3	Unacceptable	Unacceptable	Unacceptable	Borderline	Acceptable
Internal chills	G1	Unacceptable	Unacceptable	Borderline	Acceptable	Acceptable
	G2	Unacceptable	Unacceptable	Unacceptable	Borderline	Acceptable
	G3	Unacceptable	Unacceptable	Unacceptable	Unacceptable	Borderline
	G4	Unacceptable	Unacceptable	Unacceptable	Unacceptable	Unacceptable

[a] Acceptable only when angle between the defect and the direction of the principal stress is not greater than 20°.

ical tests, and service reports and were considerably more severe. The illustrations consisted of two 25 by 30-cm films and with three groupings: (1) absolute rejections (cracks, cold shuts, shrinkage cavity); (2) discrete defects (inclusions, stress segregation, and pipe); and (3) dispersed defects (porosity and microshrinkage). Two degrees of severity levels were illustrated in the discrete and dispersed categories to represent the acceptance levels for critical and noncritical areas [6].

During the early 1950s, the industry began to widely recognize the value of reference radiograph standards and sought a means of collectively developing additional standards. Around 1950, ASTM Committee E7 established Subcommittee E7.02 on Reference Radiographs and assigned it the responsibility of developing reference radiographic standards where the need was proven and sufficient user and producer interest could provide the necessary hardware. The first act of E7.02 was to adopt in 1952 the Radiographic Standards for Steel Castings from the U.S. Bureau of Ships. In 1953, a second document, E 98, was adopted, "Reference Radiographs for the Inspection of Aluminum and Magnesium Castings." In 1955, the release of

ASTM E 99–55 provided a set of 35 reference radiographs illustrating common and uncommon discontinuities in steel metal arc welds.

Early sets of reference radiographs, including the early ASTM reference radiograph documents, consisted of collections of production radiographs selected to illustrate the desired conditions and severity levels. All copies produced for sale and distribution were made by duplicating the original radiographs. Since the original hardware from which these radiographs were made no longer existed, all copies of the reference radiographs were made by film duplication. Experience with these reference radiographs pointed out some of the problems of this approach. First, the original radiographs may deteriorate over time and the production of copies of the original quality becomes impossible. Second, as radiographic sources and practices change, it is not possible to produce new reference images which are representative of these sources. As a result, current ASTM reference radiograph documents are backed by hardware. That is, all current ASTM reference radiograph documents are made using components intentionally fabricated to contain specific flaw types and sizes. These components are then carefully saved so that new original radiographs can be made as the need arises.

Document development in E7.02 began in earnest in the early 1960s. Many of the current reference radiograph documents were developed during that decade. These reference radiograph documents were developed to purpose. Rather than selecting sample radiographs from existing sources, special hardware was created to intentionally contain the desired flaw types and severity levels. Once a standard based on this hardware is established, the hardware is securely stored. Thus, new original radiographs can be made should the originals become damaged. Furthermore, superior documents can frequently be made by shooting new original radiographs rather than duplicating master radiographs. This approach was followed both for the development of new documents and for the replacements of the early ASTM reference radiograph documents.

Duplication of Reference Radiograph Documents

Development of a reference radiograph document actually involves the development of two separate documents. One of these is the text document which describes the makeup and intended use of the reference radiograph films. The other is the set of reference radiograph films. The films form an adjunct to the text document. The text document is published as part of the ASTM Book of Standards, and the adjunct is available for sale separately. Control over the content of the text document is maintained in the same way that all ASTM standards are controlled. This includes a multilevel balloting process, editorial review, printing, and distribution. The film adjuncts, on the other hand, require a somewhat different control system. During the development process and balloting of the text document, the proposed film set is made available for numerous reviews by ASTM and industry participants. Refinements are made in the film set until a consensus is reached on the suitability of the film images. Various forms of documentation may be used to validate the graded series, including destructive tests of components with similar conditions and computer analysis of the discontinuity images. Once the film set is accepted, it becomes a prototype for the adjunct materials. Additional sets must be produced to provide inventory for sales of these documents.

The faithful reproduction of a prototype document offers numerous challenges. Various methods have been evaluated. One early approach involved the use of a photographic intermediate transparency produced using either long focal length lenses or contact printing techniques. The intermediate transparency was then used to print multiple copies of the original radiographic transparencies [7]. This process provided a means of duplicating the radiographs but frequently resulted in variations in contrast and graininess that at times caused difficulties in reproducing the finer discontinuity grades.

Once made-to-purpose hardware became available, many of these film adjuncts could be produced by preparing original radiographs of the actual hardware. In many cases, complete graded sets of discontinuities which are illustrated on a single page of the adjunct are assembled into a welded block and are radiographed as a set onto a single film. Shims may be added to some of the discontinuity sections to produce uniform film density on the resulting radiograph. While this approach is frequently effective for many of the light alloys or thin sections, very long exposure times make this approach impractical for many of the thicker sections.

Another common approach utilizes direct positive duplication film. Early versions of this type of film provided very good representation of the contrast and graininess of the original radiograph; however, the suitable density range for these films tended to be too low. Neutral density films were added to bring the average density up to a suitable level [7]. Current duplication films do not suffer from this limitation, and normal densities can be accurately reproduced. Many of the current reference radiograph adjunct documents are now produced using a direct duplication process.

Through the 1960s and 1970s, many of the current reference radiograph documents were developed and sales began to escalate. Therefore, the need to produce large numbers of reference radiograph sets which faithfully represented the originals became critical. The need for a quality control system to insure the faithfulness of these images became apparent. Around 1978, Subcommittee E7.93 on Illustration Monitoring was formed to provide this quality control system for the reference radiograph adjunct documents. E7.93 review teams now examine every reference radiograph page produced for sale to the public for faithful rendering of the quality, contrast, density, and appearance of the prototype masters. E7.93 also provides the final acceptance of the prototype master. This practice of reviewing every reference radiograph image has resulted in extremely consistent images being produced. The reproduction of reference radiographs has improved through the years to the point where, today, every set of a particular reference radiograph document is virtually identical.

Present Status of Reference Radiographs

ASTM currently maintains 13 reference radiograph documents incorporating 23 volumes of adjunct films covering a wide variety of casting materials and steel weldments. An additional document illustrates, in line drawing format, various defects and irregularities in semiconductors and electronic components. Great care has been taken to illustrate natural discontinuities, which represent common defects in welds and castings. In many cases the severity levels illustrated were selected based on demonstrated reduction of engineering properties using destructive testing of similar components [8]. This does not imply that in all cases a Grade 2 discontinuity will result in, say, a 10% reduction in properties. It should be clear that no such relationship can exist, since component design, defect location, loading conditions, and other factors will affect the actual performance of the object. In fact, it should be recognized that each defect type should be considered separately for each component or structural zone of a component [9]. For example, a particular weldment may be able to tolerate Grade 4 Fine Scattered Porosity, while Grade 2 incomplete penetration would be unacceptable. It must be realized that the interpretation of many radiographic indications continues to require the trained eye of an experienced radiographer. For most discontinuity types, there is an infinite array of possible representations and any set of reference radiographs can only hope to show a selection of representative illustrations. Sound judgment and experience must still be applied to the interpretation of conditions in production hardware. Factors such as the size distribution, relative spacing, shape, and density of indications such as porosity must be carefully considered and weighed since it is unlikely that any component will exhibit a condition which exactly matches the reference radiograph condition.

TABLE 2—*Current reference radiograph documents.*

Document No. and Year of Original Issue	Material and Fabrication	Section Thickness, mm (in.)	Discontinuity Types Illustrated
E 155 (1960), 2 volumes	Aluminum and magnesium castings	6 to 19 mm (¼ to ¾ in.)	*Aluminum:* Gas holes; gas porosity, round and elongated; shrinkage cavity; shrinkage; foreign material *Magnesium:* Gas holes; microshrinkage, sponge & feathery; foreign material; reacted sand inclusions; segregation, eutectic and gravity *Ungraded illustrations:* Pipe shrink, flow line, hot tear, oxide inclusion
E 186 (1962), 3 volumes by energy level	Steel castings, heavy walled	51 to 114 mm (2 to 4 ½ in.)	*Graded:* Gas porosity, sand and slag inclusions, shrinkage *Illustrations:* crack, hot tear, insert
E 192 (1962)	Investment steel castings	Up to 25 mm (1 in.)	*Gas holes:* shrinkage, 4 types; foreign materials; hot tears; cold cracks; cold shut; misruns; core shift; mold buckle; diffraction patterns
E 242 (1964)	Various materials appearances as parameters are changed	Up to 152 mm (6 in.)	150 kVp, 250 kVp, 1MV, 2 MV, 10 MV, 15 MV, Iridium-192, Cobalt-60, Radium-226
E 272 (1965)	Copper-based and nickel-based castings	Up to 51 mm (2 in.) and 51 to 152 mm (2 to 6 in.)	Gas porosity; sand inclusions; dross inclusions; shrinkage, 3 types
E 280 (1965), 2 volumes by source energy	Steel castings, heavy walled	114 to 305 mm (4½ to 12 in.)	Gas porosity; sand and slag inclusions; shrinkage, 3 types; hot tear; insert
E 310 (1966)	Tin bronze castings	Up to 51 mm (2 in.)	Gas porosity; shrinkage, 2 types; sand inclusions; hot tear; insert
E 390 (1969), 3 volumes by thickness	Steel fusion welds	V1: up to 6 mm (up to ¼ in.) V2: 6 to 76 mm (¼ to 3 in.) V3: 76 to 203 (3 to 8 in.)	Fine scattered porosity, coarse scattered porosity, linear porosity, slag inclusions, tungsten inclusions, incomplete penetration, lack of fusion, worm hole porosity, burn through, icicles, cracks, undercut
E 431 (1971)	Semiconductors and related devices	N/A	No radiograph transparencies, line drawing illustrations of various component irregularities
E 446 (1972), 3 volumes by source energy	Steel castings	Up to 51 mm (2 in.)	Gas porosity; sand and slag inclusions; shrinkage, 4 types; crack; hot tear; insert; mottling
E 505 (1974)	Aluminum and magnesium die castings	Up to 25 mm (1 in.)	Porosity, cold fill, shrinkage, foreign material
E 689 (1979)	Ductile iron castings	Up to 304 mm (12 in.)	Use E 186, E 280, and E 446 for illustrations

TABLE 2—*Continued*

Document No. and Year of Original Issue	Material and Fabrication	Section Thickness, mm (in.)	Discontinuity Types Illustrated
E 802 (1982), 3 volumes by source energy	Gray iron castings	Up to 114 mm (4½ in.)	Centerline shrinkage, (Use with E 186 and/or E 446 for other discontinuities)
E 1320 (1990) 2 volumes by thickness	Titanium castings	V1: up to 25 mm (up to 1 in.) V2: 25 to 51 (1 to 2 in.)	Clustered gas holes, scattered gas holes, shrinkage cavity, centerline shrinkage, inclusions

The currently available ASTM reference radiograph documents and the materials and forms covered are summarized in Table 2. The newest of these is for titanium castings and just became available in 1991. A new document for aluminum weldments [10] and an additional volume covering aluminum and magnesium castings 25 to 63.5 mm in thickness are currently being prepared.

Assembly of the reference radiograph images for the aluminum welds document is nearly complete. Over 200 welds have been made from which approximately 30 have been selected for use in the reference radiograph images. This represents approximately 85 to 90% completion of the weld preparation. The welds are in 3 and 12-mm aluminum plates to represent the range of 0 to 19 mm. The planned welds are summarized in Table 3. The completed set was made available for review at the June 1991 meeting of ASTM Committee E7.

These reference radiographs are being subjected to computer analysis of the weld images. The weld radiographs are placed on a light table and scanned into a computer digitized image. Image-processing tools can be used to locate and identify defect areas on the film. These defect areas can then be analyzed as to size, number, distribution, etc. The images of scattered and linear porosity are now being plotted based on projected area fraction of the void space versus grade level. This is proving to correlate very well with the grade selection for the various poros-

TABLE 3—*Status of aluminum weld reference radiographs.*

Discontinuity Type	Thickness Range, 9.5 to 19-mm Grades/Levels	Thickness Range, 0 to 9.5-mm Grades/Levels
Fine scattered porosity	5	5
Coarse scattered porosity	5	. . .
Linear porosity	5	5
Inadequate penetration	2	2
Tungsten inclusions	2	. . .[a]
Cracks	2	2
Undercut	1	1
Crater cracks	1	. . .[a]
Clustered porosity	1	. . .[a]

[a]Weld radiographs from the thicker set will be used for illustrations in both thickness ranges.

ity types and thicknesses both in terms of trend and level. For example, the curves for fine and coarse porosity in the 12-mm samples and for fine porosity in the 3-mm samples all showed a similar shape and all represented approximately 3% of the projected area for Grade 4.

ASTM Subcommittee E7.02 on Reference Radiographs is also involved in some peripheral areas relating to radiographic imaging and interpretation. The subcommittee is responsible for E 592: Guide to Obtainable ASTM Equivalent Penetrameter Sensitivity for Radiography of Steel Plates ¼ to 2 in. (6 to 51 mm) Thick with X-Rays and 1 to 6 in. (25 to 152 mm) Thick With Cobalt-60. This document describes and illustrates the influence of source type and energy, film type, and material thickness on the equivalent penetrameter sensitivity.

The subcommittee is also currently working on the development of a standard for evaluating the visual acuity of radiographic inspectors. Many inspectors are routinely evaluated for color perception, but the ability of the interpreter to perceive subtle radiographic detail under various conditions is generally not determined. The focus of the work is to develop a test system and procedure which will rapidly determine the ability of the interpreter to perceive subtle radiographic indications. The test could be used to evaluate not only the performance of an individual interpreter, but also the suitability of lighting in an interpretation facility or the effects of fatigue on operator performance.

The visual acuity work is based on earlier work conducted by the National Bureau of Standards (now the National Institute of Standards and Technology) [11]. In this effort, a series of test images is being produced by radiographing test blocks to varying parameters. Each test image consists of a 5 by 5-cm image of uniform density except for a small line at one of five locations and four orientations. The test set contains lines of various contrast levels and sharpnesses. The time required to complete the reading for the full set is typically 5 to 10 min.

ASTM reference radiograph documents are recognized worldwide as a source of visually communicated references to appearances and severities of radiographically detected discontinuities. Some 600 to 800 volumes of adjunct materials are sold each year throughout the world, particularly in the Americas, Europe, and Asia. The ASTM reference radiograph documents do not attempt to identify the grades which should be selected for various applications, but rather provide a range of severity levels which may be selected and negotiated as required between purchaser and supplier. This approach provides the maximum flexibility in the use of these documents and avoids the false implication that a particular grade of discontinuity will always result in the same structural significance regardless of the intended application, environment, and loading of the inspected component.

Future Directions for Reference Radiographs

The future for reference radiograph documents holds the promise of an expanding array of materials, including composites, ceramics, and other new materials. The early application of radiographic techniques to these materials involves the case by case interpretation and determination of the significance of any indications. This approach parallels the early use of radiographic methods on weldments and castings. As acceptance criteria are developed by individual users, those acceptance criteria become codified in internal documents [12,13]. As more people begin to use these materials and radiographic inspection is more uniformly applied the need may come for industry consensus standard reference radiograph documents. Where sufficient interest in such a document is evident, this development activity will be aggressively pursued.

Another technology which promises to present new challenges to the development of reference radiologic images involves a variety of new radiologic techniques, particularly including conventional and microfocus radioscopic imaging techniques. These techniques are becoming widely used in a variety of industries and are codified in a variety of ASTM standards [14–16]. Radioscopic imaging techniques differ from film radiographic techniques in several very important ways. First is the difference in the imaging tool itself. X-ray film offers extremely high resolution and large density latitude. Video images are typically made up of far fewer image elements and have a narrower instantaneous density range. On the other hand, video radioscopic images can be rapidly updated to provide real time imaging and can be interfaced with sophisticated image processing systems to extend the image representation over a wide range of densities and applications.

Two primary factors must be addressed for the development of suitable parallels to the film images for use with radioscopic systems. First, the technical makeup of the image must be established. With film systems, common practice calls for the object to be placed in contact with the film cassette. This results in an image which is essentially true size to the object radiographed. With radioscopic systems, significant geometric magnification is frequently used. In fact, depending on the particular application and system being used, geometric magnifications of 1.5 to ×50 or more are common. Further, the video displays used vary from relatively low-resolution systems producing less than 500 video lines, to systems which produce in excess of 2000 lines. It will be a challenging task to establish a practical approach to producing standard reference images which will satisfy all of these conditions.

The second major consideration involves the format of the presentation of the reference images. Most real time radioscopic systems include some recording and playback capability for image archiving. A system may include one or more of the following: video tape recorders, optical disk systems, and magnetic disk systems. Each of these approaches involves the storage of a representation of the image which can be reconstructed into an actual image at a later point in time. If this approach is to be used for the reference images, the reconstruction approach becomes a part of the reference image and will have to be taken into consideration. The faithfulness with which the actual image is reconstructed could potentially vary from one system to another. Another approach currently being considered is the development of a film transparency which illustrates the reference image in a manner which simulates a video display. The reference image in this case would be displayed on a separate backlit film viewer to produce the appearance of a video screen. This approach has the advantage that the reference images presented at different locations would be more consistent, but has the disadvantage that the presented image may not adequately simulate the appearance for the particular radioscopic system in use. These factors will have to be carefully considered and studied as development of these standards is considered.

Another significant advance in the technology of radiology involves the use of image processing systems to begin to automatically interpret radiological images. Some of the early work in this area has already been accomplished [17–20]. The advantages of image processing and analysis can also be applied to film radiographs by first electronically scanning the film image into a computer data base [21]. These systems involve the use of image processing technology to analyze the image and extract discontinuity information from the component image information. In various systems this may be accomplished using a reference image of a defect free part or using computer algorithms to identify discontinuity images. The output of these systems may be an automatic acceptance or rejection of the inspected component or it may be a modified image in which discontinuities are highlighted to the operator for grading and acceptance. As this type of approach is developed further, the acceptance levels represented by the reference radiographs may need to be codified into rules which can be used by the computer analysis systems to grade discontinuities. This clearly represents a major shift in the concept of reference images and will require careful study and consideration.

Another radiologic modality, computed tomography, is also becoming widely used in a variety of industries. Activity is underway within ASTM Subcommittee E7.01 to develop standards for computed tomography practice. The cross-sectional imaging capability of this modality offers still more considerations for the standardized communication and specification of acceptance levels for various discontinuity conditions.

The adaptation of reference radiograph documents for radioscopic applications offers a variety of challenges in the preservation and control of image quality and presentation so that the reference images provide correct representations of the specified condition. Further application of reference radiograph flaw standards in systems which provide automated interpretation will require consideration of a host of factors. Every effort will continue to be made to provide images which accurately communicate the applicable acceptance levels to radiographic interpreters in a consistent manner. In some cases, new concepts may need to be developed for the specification and communication of acceptance levels.

References

[1] Briggs, C. W., "Developments in Gamma Ray Radiography 1928–41," *Industrial Radiography*, Vol. 1, No. 1, 1942, p. 7–10.
[2] Clason, C. B., "X-Ray Testing of Welds," *Industrial Radiography*, Vol. 1, No. 2, Fall 1942, pp. 14–20.
[3] Frear, C. L., "Radiographic Specifications and Standards for Naval Materials," *Industrial Radiography*, Vol. 2, No. 4, Spring 1944, pp. 33–36.
[4] Busk, R. S., "A Correlation of the Mechanical Properties and Radiographic Appearance of Magnesium Alloy Castings," *Industrial Radiography*, Vol. 2, No. 3, Winter 1943–44, pp. 33–37.
[5] Wyle, F. S., "The Use of Static Tests as a Method of Determining the Radiographic Classification of Castings," *Industrial Radiography*, Vol. 3, No. 4, Spring 1945, pp. 13–30.
[6] Pierce, J. J., "Radiographic Specifications; Their Nature, Purpose, and Current Revisions," *Industrial Radiography*, Vol. 5, No. 4, Spring 1947, pp. 21–23, 50.
[7] Goldspiel, S., "Development of Radiographic Standards for Castings," *Materials Research and Standards*, Vol. 9, No. 7, July 1969, pp. 13–20.
[8] Goldspiel, S., "Status of Reference Radiographs," *Nondestructive Testing Standards—A Review*, ASTM STP 624, H. Berger, Ed., American Society for Testing and Materials, Philadelphia, 1977, pp. 115–128.
[9] Halmshaw, R., "Interpretation of Radiographs," *Industrial Radiology: Theory and Practice*, Applied Science Publishers, London, 1982, pp. 243–269.
[10] Berger, H., Polansky, D., and Criscuolo, E. L., "A Data Bank for Graded Reference Radiographs of Aluminum Welds," *Non-Destructive Testing (Proc. 12th World Conference)*, J. Boogaard and G. M. van Dijk, Eds., Elsevier Science Publishers, Amsterdam, 1989, pp. 86–89.
[11] Yonemura, G. T., "Visual Acuity Testing of Radiographic Inspectors in Nondestructive Inspection," technical report, National Bureau of Standards, Center for Building Technology, Washington, DC, 1981.
[12] P.S. 21206.3: Radiographic Inspection of Composite Honeycomb Assemblies and Structures, McDonnell Douglas, St. Louis.
[13] P.S. 21233: Nondestructive Testing of Honeycomb Assemblies and Composite Structures, McDonnell Douglas, St. Louis.
[14] E 1000: Guide for Radioscopic Real-Time Imaging, American Society for Testing and Materials, Philadelphia, PA.
[15] E 1255: Practice for Radioscopic Real-Time Examination, American Society for Testing and Materials, Philadelphia, PA.
[16] E 1411: Practice for Qualification of Radioscopic Systems, American Society for Testing and Materials, Philadelphia, PA.
[17] Munro, J. J., et al., "Weld Inspection by Real-Time Radioscopy," *Materials Evaluation*, Vol. 45, No. 11, Nov. 1987, pp. 1303–1309.
[18] Daum, W., et al., "Real Time Evaluation of Weld Radiographs by Digital Image Processing," *Nondestructive Testing*, J. M. Farley and R. W. Nichols, Eds., Pergamon Press, Oxford, 1988, pp. 1568–1574.

[*19*] Eckelt, B., et al., "Use of Automatic Image Processing for Monitoring of Welding Processes and Weld Inspection," *Non-Destructive Testing (Proceedings of the 12th World Conference)*, J. Boogaard and G. M. van Dijk, Eds., Elsevier Science Publishers B.V., Amsterdam, 1989, pp. 37–41.

[*20*] Rose, P., et al., "Digital Image Data Base for Flaw Classification in Radiographs," *Nondestructive Testing*, J. M. Farley and R. W. Nichols, Eds., Pergamon Press, Oxford, 1988, pp. 1564–1567.

[*21*] Graeme, W. A., Eizember, A. C. and Douglass, J., "Digital Image Analysis of Nondestructive Testing Radiographs," *Materials Evaluation*, Vol. 48, No. 2, Feb. 1990, pp. 117–120.

Jack S. Brenizer[1]

Current and Future Neutron Radiologic NDT Standards

REFERENCE: Brenizer, J. S., **"Current and Future Neutron Radiologic NDT Standards,"** *Nondestructive Testing Standards—Present and Future, ASTM STP 1151,* H. Berger and L. Mordfin, Eds., American Society for Testing and Materials, Philadelphia, 1992, pp. 34–40.

ABSTRACT: Two groups are active in producing standards which are used in neutron radiology (NR) activities within the United States. On the national level, ASTM has three standards which address the neutron radiography method: ASTM Practices for Thermal Neutron Radiography of Materials (E 748), ASTM Method for Determining Image Quality in Direct Thermal Neutron Radiographic Testing (E 545), and ASTM Method for Determining the L/D Ratio of Neutron Radiography Beams (E 803). The International Organization for Standardization (ISO) has a working group which is developing three standards that also address the neutron radiography method. No ASTM or ISO standards exist for the neutron radioscopic NDT method.

Future ASTM standards will address the neutron radioscopic method, neutron radiographic dimensioning, and neutron radiologic system characterization. It is expected that similar efforts will be undertaken in ISO. Given the relatively small community providing neutron radiologic NDT services, international cooperation already in place will most likely continue to grow. This paper reviews current and future trends in NR standards and the impact of these standards on the world market.

KEY WORDS: standards, neutron radiography, neutron radioscopy, neutron radiology, nondestructive testing

Images made with neutrons have been used in nondestructive testing (NDT) since the early 1960s for a wide variety of industrial and research applications. Several authors have written articles and handbooks reviewing the technique of making radiographic images using neutrons [1,2,3,4]. The most common source of neutrons for this NDT method has been and remains a research fission reactor. This has limited the applications and users of neutron radiography due to the small number of reactor facilities available and the high cost of operation relative to the number and operating costs of X-ray radiographic systems. Thus, alternative NDT methods are often used even when the alternative technique yields less information.

Neutron radiography creates an image which looks like an X-ray radiograph, but the differences between neutron and X-ray interaction mechanisms produce images which contain completely different information. While X-ray attenuation is directly dependent on atomic number, neutrons are efficiently attenuated by only a few specific elements. This is because X-rays interact with the electron cloud of an atom; the more electrons, the greater the attenuation. Neutrons, on the other hand, interact with the nucleus, and this interaction is highly dependent on nuclear mass and structure. Boron-10 and gadolinium are good attenuators of thermal neutrons due to their large absorption cross sections, whereas hydrogen is a good attenuator by virtue of its large scattering cross section. As a result, organic materials and water

[1] Associate professor of Nuclear Engineering, University of Virginia, Department of Nuclear Engineering and Engineering Physics, Charlottesville, VA 22903-2442.

are clearly imaged in most neutron radiographs because of their high hydrogen content, while many structural materials such as aluminum or steel are largely transparent.

As early as 1969, it was recognized that some standardization was necessary in the neutron radiography field. Over 15 papers on this subject have been presented at the three world conferences on neutron radiography. Haskins has presented two reviews of neutron radiography standards in the United States, and a later paper by Newacheck and Tsukimura updates these earlier papers [5,6,7]. The American Society for Testing and Materials (ASTM) is the principal group within the United States concerned with developing standards for neutron imaging. Personnel qualification of neutron radiographers is under the American Society of Nondestructive Testing (ASNT). Work on developing standards for neutron radiography began in 1971, and the first ASTM standard directly dealing with neutron radiography was published in 1975. By 1981, two additional neutron radiography standards had been published. Although these standards have gone through some revision, the basic concepts have remained unchanged.

The reference to neutron imaging as radiography resulted from its similarity to the X-ray radiographic technique. The development of electronic imaging systems in the mid-1970s began a new field, first referred to as real-time neutron radiography. Most commonly, the neutron attenuation intensity pattern impinges on a scintillating phosphor which, after intensification, is viewed with a video camera. The typical frame rates were sufficiently fast to permit observation of dynamic motion without blurring, hence the name real-time.

By the mid-1980s, the ASTM E7.01 Radiology (X and gamma) method and E7.05 Radiology (neutron) method subcommittees recognized the need to more clearly define the term *real-time radiography*. The terms *radiology*, *radioscopy*, and *radiography* have been defined and can be found in ASTM Terminology for Nondestructive Examinations (E 1316). Radiography is now used to describe techniques which produce a permanent visible image on a recording medium, usually film. Radioscopy is used to refer to techniques which use electronic production of a radiological image that very closely follows the changes in the image with respect to time. Radiology is defined as the science and application of penetrating radiations.

Radioscopic systems added a new dimension to the neutron radiology NDT methods. They also permitted neutron imaging with shorter exposure times. This has led to the use of isotopic and accelerator neutron sources for neutron imaging. The concept of portability and the observation of dynamic motion has renewed interest in the use of neutron radiology in NDT.

The development of more exotic materials, especially composites, and the increased use of aluminum as a structural material, coupled with the increasing availability of portable systems, suggest that neutron radiology (NR) will continue to be a valuable NDT tool. The NR community has been quite small, with most NR groups based at reactor facilities. As the use of NR NDT methods grow, new systems will be purchased and new groups will be offering NR services. The need for uniform standards for system characterization and NDT radiologic practice will grow with the expanded utilization of the method. This paper reviews the current and future trends in NR standards and the impact of these standards on the world market.

Current Neutron Radiologic Standards

Two groups, ASTM and ISO, are active in producing standards which are used in neutron radiologic (NR) NDT activities within the United States. On the national level, ASTM has three standards, all of which address neutron radiography. The emphasis on radiography is understandable since most neutron radiology applications to date have employed either direct or indirect neutron radiographic techniques, specifically with thermal neutrons. These NDT methods have been used to examine a wide variety of objects ranging from jet engine turbine blades to spent nuclear fuel.

ASTM Practices for Thermal Neutron Radiography of Materials (E 748) provides a good introduction to the neutron radiographic technique. This document was intended to be somewhat tutorial in nature, describing the thermal neutron radiographic method and required facilities and equipment. A description of neutron sources, beam filters, and collimators is presented, along with a list of background references. The use of conversion screens and film cassettes for both the direct and indirect imaging methods is discussed in detail. The standard also addresses the materials and general applications for which the neutron radiographic technique is appropriate.

The ASTM Method for Determining Image Quality in Direct Thermal Neutron Radiographic Testing (E 545) is widely used by neutron radiography practitioners both in the United States and in other countries. This standard is used to determine the relative overall quality of the neutron radiographic images produced. It is not intended to be used for controlling the acceptability of quality of materials and components. The judgment of the radiograph's quality is based upon the evaluation of images obtained from two different indicators that are exposed simultaneously with or under exactly the same conditions as those used to examine a test object.

The first device, the beam purity indicator (BPI), is used to obtain a quantitative determination of radiographic quality. Construction of the device is straightforward. It consists of a TFE-fluorocarbon block containing two boron nitride disks, two lead disks, and two cadmium wires. A qualitative determination of a facility's neutron beam can be obtained from a visual inspection of the BPI's radiographic image. Densitometric measurements taken at specified locations on the BPI's image are used to calculate the radiographic contrast, low-energy photon and pair-production contributions to the image, the image unsharpness, and information on the film and processing quality. A specific procedure for the densitometric measurements and calculations is given in the standard.

The second device, the sensitivity indicator (SI), is used to qualitatively determine the sensitivity of detail visible on the neutron radiograph. The device is made from acrylic, aluminum, and lead components. Solid aluminum shims are used to create low-density gaps in the SI's radiographic image. Holes in acrylic shims also create areas of low sample density visible on the film image. Visual inspection of the SI's image is used to determine the smallest gaps and the number of consecutive holes detected. Subjective information regarding the level of detrimental gamma photon exposure is also available.

The information obtained from the BPI and SI radiographic images is then used to determine a neutron radiographic category for the facility. The BPI yields quantitative information concerning neutron beam and imaging system parameters. It can also be used as a daily check on consistency of radiographic quality. Likewise, the SI can be used as a check on sensitivity.

The ASTM Standard Method for Determining the L/D Ratio of Neutron Radiography Beams (E 803) is also widely used by neutron radiography practitioners. It provides an experimental technique to determine the ratio of the effective collimator length (L) to effective collimator entrance aperture diameter (D). This is different from the simple ratio of the physical collimator length and aperture diameter, since neutrons scattered off both the collimator and shielding walls affect the L/D ratio.

The E 803 method involves examining the radiographic image of a no-umbra device to determine the point where the umbral shadow disappears. The device consists of a U-shaped aluminum channel with a series of parallel V grooves at specified intervals along its length. Each groove contains a thin cadmium wire of known diameter. The device is placed on the film cassette at a 45° angle. A single device will allow determination of L/D ratios up to 150; higher L/D values can be measured by adding a second device similar to the first.

The L/D ratio is evaluated using the following procedure. A radiograph of the no-umbra device is made such that the background density is approximately 2.5. The film is then analyzed using one of three alternate methods. In the first, a visual analysis is used to determine where the umbra disappears. The ratio of the rod position with zero umbral shadow width to the rod diameter is equal to the effective L/D ratio. The second alternate method involves the use of a microdensitometer to determine the zero umbral shadow location. This method should only be used with L/D ratios up to several hundred. A third alternative method requires the use of a microdensitometer to examine the individual shadow waveforms to determine the width of two different umbral images. Subsequent calculations using these values yields the L/D ratio. The third method is useful in determining both high and low L/D ratios.

Neutron radiology standards are also being developed at the international level. The International Organization for Standardization (ISO) appointed a working group, ISO/TC 135/SC 5/WG 4 "Thermal Neutron Radiography," in April, 1988. Ten countries have representatives participating in the working group as either members or observers. The working group has met on six occasions to discuss and draft three work items. Work Item 5.5 "Non-destructive testing; Thermal neutron radiographic testing; General Principles" is based on ASTM E 748. The draft was balloted under SC 5 last year and after editorial revisions has been sent back to SC 5.

Work Item 5.6 "Non-destructive testing; Thermal neutron radiographic testing; Determination of Beam L/D ratio" has been reviewed and rewritten several times since the original submission of ASTM E 803 as a working draft. Several alternative L/D measurement methods and devices have been considered. It now appears that the general consensus of the working group is that the ASTM E 803 no-umbra device and method is the best of several proposed techniques. A final draft is now being prepared.

ASTM E 545 was submitted as the first draft for Work Item 5.7 "Non-destructive testing; Thermal neutron radiographic testing; Determination of image quality." This draft was not immediately accepted due to disagreements between the European Neutron Radiography Working Group's (NRWG's) round robin test results and those obtained in U.S. round robin tests [8,9]. Discussions between ASTM E 7.05 subcommittee and NRWG representatives led to an understanding of the differences. The different test results were caused by inconsistencies in SI device manufacture. After these were rectified, similar test results were obtained by both groups. Several alternative image quality devices were considered by the ISO working group. Again, it appears that the ISO working group is preparing to base its draft of Work Item 5.7 on the ASTM SI and BPI devices with improved material and manufacturing specifications.

At the present time, no ISO standard for neutron radiography exists. It is anticipated that the "Non-destructive testing; Thermal neutron radiographic testing; General Principles" document will be the first ISO standard in this area. Three ASTM neutron radiographic standards are available. However, no ASTM or ISO standards currently exist for neutron radioscopy.

Future Neutron Radiologic Standards

The ASTM E 545, E 748, and E 803 standards will continue to be used for the foreseeable future. These standards are under periodic review and are revised to reflect both improvements in the procedures, methods, and devices used, as well as to correct incomplete, inaccurate, or outdated data. New ASTM standards currently being drafted include a standard on neutron radioscopic practice, a method for neutron radiographic dimensioning, and a method for neutron radiologic system characterization. The most difficult of these is the method for neutron radiologic system characterization.

A standardized approach for characterizing radiographic and radioscopic systems is desirable and will aid in the interpretation of NR images and will permit an intercomparison of images obtained using different facilities. This concept is not new, and an approach was described by Bayon and Laporte [10]. The characterization should take into consideration both reactor or other neutron sources. Parameters such as beam divergence, beam uniformity, beam area, beam orientation, image system performance, dimensional calibration (and uniformity), and contrast sensitivity are all important in understanding and interpreting NR images. Most of these parameters are not addressed in the existing standards. Several characterization devices and parameters, such as the cadmium ratio, modulation transfer functions (MTF), neutron flux, neutron-to-gamma ratio, resolution, and the values measured using the current ASTM BPI and SI devices are often used or referred to when describing an NR facility or system, but they are not determined in a standardized manner.

The E7.05 subcommittee and the ISO working group have considered several proposals for system characterization. The general consensus at both the national and international level is that radiographic and radioscopic systems are too complex to adequately characterize with a single parameter or device. The favored approach is one which will involve the establishment of standard methods for the measurement of many of the beam, imaging, and facility parameters mentioned above, coupled with a guide for interpreting the parameter's overall impact on an NR image. Standardization of measurement and interpretation is important since some of the parameters are interrelated. For example, values obtained for the cadmium ratio can be affected by beam filters and neutron energy spectrum.

Many existing standards can be utilized. Some standards, such as the ASTM Standard Method for Determining Neutron Flux, Fluence, and Spectra by Radioactivation Techniques (E 261), are well established but not currently used in describing NR system characteristics. The BPI and SI devices work well for radiography, but either these devices need to be revised or alternative devices developed for use with radioscopic systems. Some standard devices and techniques used in gamma and X-ray radiology can possibly be adapted from NR characterization. However, for many of the desirable characterization parameters, relevant or existing standards simply do not exist. A logical approach would be to establish a priority for each of the characterization parameters, develop a method for measuring the parameter, proceed to round robin testing of the method, and then proceed to draft and approve a standard method through ASTM. After a sufficient number of standard methods have been adopted, an overall system characterization document can be established.

It is expected that similar efforts will be undertaken in ISO. New devices are continually being proposed by various NR groups at both national and international NR facilities. Both the ASTM and ISO/TC 135/SC 5/WG 4 evaluate every proposed method or device. For example, at recent ISO/TC 135/SC 5/WG 4 and ASTM E7.05 meetings, a total of seven new standard methods or devices for measuring characterization parameters were proposed. Those with strong technical documentation or experimental support or those endorsed by national or international groups are evaluated by round robin testing. Some proposals have been accepted for consensus review and approval, some have been rejected, and several are still under investigation. Many will eventually become ASTM or ISO standards.

World Impact of NR Standards

The NR community is already very international in nature. Most technical meetings are international gatherings. Observers and members from ten different countries participate in the ISO working group. The international demand for NR NDT services already in place will most likely continue to grow. Given the relative expense of NR NDT services, comparison of services offered and the quality of services delivered will become increasingly competitive. Standardization of important terms and parameters used to specify the characteristics of a facility's NR system and capabilities is an important factor for several reasons.

Standardization provides a basis for prospective customers to make a judgment of which facility is best suited to their particular NDT needs. This may at first seem to put some facilities at a disadvantage, but in NR as in most techniques, what is most important in one case may be detrimental in another. If the NR method is to continue to grow in its application, customers must be able to get good results at a reasonable cost. Using the facility best suited to the project's needs and obtaining good results serves to increase the number of future applications for NR methods.

Standardization provides a common language among the customers and practitioners of the NR method. This is important as it serves to provide a clear picture of the NR NDT method to potential national and international customers and also aids in the establishment of agreements for international NR services. Neutron radiologic facilities are expensive to build, especially if a reactor is needed, and they must comply with complex governmental regulations. While this prevents many from establishing their own facility, it promotes the use of international services. Industrial applications which require NR NDT have been limited in number primarily due to availability and cost. However, applications requiring NR techniques are often difficult or impossible problems to solve using other NDT techniques, and the integrity of the part or assembly is critical to an operation's success. Consequently, NR NDT research or service projects frequently involve expenditure of large quantities of money.

Standardization provides a method to monitor the performance of an NR system over time. It also provides a means to evaluate changes made in system components and their effect of the change on an NR image. This is of particular importance to system developers, vendors, and those installing new or updating existing systems. If the number of NR facilities grows, this will be an important role for the system characterization standards, even on an international basis as systems and system components are sold around the world.

Conclusions

The existing ASTM NR standards have been widely used in both the United States and in other countries. Although their accuracy has been questioned and other alternative methods of measuring similar parameters have often been proposed, the ASTM standards' validity and usefulness have always been confirmed by a consensus of those practicing NR. The need for additional ASTM NR standards has been apparent for several years, but since the number of facilities using NR techniques has remained small, development of these standards has been slow. It is anticipated that the next ASTM standard adopted will be in dimensional measurement, followed by a standard practice for neutron radioscopy.

The most difficult NR standards to develop will be those used for NR system characterization. Many parameters are already in common use when describing an NR system, but they are not currently measured using a standardized method. System characterization will most likely consist of one standard which gives an overall interpretation of parameters measured according to additional separate standards. Some of these standards already exist, but most must be developed. The need for this new set of standards will become greater as more new NR systems are placed into use and older NR systems are upgraded. A standardized approach to system characterization will also be more important for those developing NR systems.

As the number of customers for NR services grows, standardization will be a valuable tool in providing a uniform language for describing NR NDT services. There is currently a good exchange of NR services and technical information at the international level. The ISO/TC 135/SC 5/WG 4, the NRWG, and the ASTM Subcommittee E7.05 have for many years had an open exchange of information and review of existing and proposed standards. This exchange has strengthened in the last five years and should continue to grow. The small size of the NR community has helped to keep this NDT technique unified without a large number of standards.

On the international level, the ISO/TC 135/SC 5/WG was established and has begun to develop a set of radiographic standards. After only three years, this working group has sent a draft standard to the subcommittee level and has reviewed two additional draft standards. While progress has been slower at the international level than at the national level, the working group has provided a good format to have all prospective ideas reviewed on the international level. It is anticipated that the ISO/TC 135/SC 5/WG will continue to work on the remaining two initial radiographic work items, but its role will be expanded to include both a radioscopic standard and alternative methods for system characterization.

The increased NR activity and expansion of the applications has generated an interest in having more standards. It is imperative that all of the groups using NR be involved in preparation of standards. Standards will only be a valuable tool if both the customers and practitioners of the NR method accept them and use them on a routine basis.

References

[1] Barton, J. P., "Neutron Radiography—An Overview," *Practical Applications of Neutron Radiography and Gaging, ASTM STP 586*, H. Berger, Ed., American Society for Testing and Materials, Philadelphia, 1976, pp. 5–19.

[2] Hawkesworth, M. R. and Walker, J., "Basic Principles of Thermal Neutron Radiography," *Neutron Radiography: Proceedings of the First World Conference*, J. P. Barton and P. von der Hardt, Eds., D. Reidel, Dordrecht, Holland, 1983, pp. 5–21.

[3] Whittemore, W. L. and Berger, H., "Physics of Neutron Radiography Using Selected Energy Neutrons," *Neutron Radiography: Proceedings of the First World Conference*, J. P. Barton and P. von der Hardt, Eds., D. Reidel, Dordrecht, Holland, 1983, pp. 23–33.

[4] von der Hardt, P. and Röttger, H., Eds., *Neutron Radiography Handbook*, D. Reidel, Dordrecht, Holland, 1981.

[5] Haskins, J. J., "ASTM Activities in Neutron Radiography," *Practical Applications of Neutron Radiography and Gaging, ASTM STP 586*, H. Berger, Ed., American Society for Testing and Materials, Philadelphia, 1976, pp. 106–113.

[6] Haskins, J. J., "Neutron Radiography Standards in the United States of America," *Neutron Radiography: Proceedings of the First World Conference*, J. P. Barton and P. von der Hardt, Eds., D. Reidel, Dordrecht, Holland, 1983, pp. 985–991.

[7] Newacheck, R. L. and Tsukiumur, R. R., "Current Status of the ASTM E 545 Image Quality Indicator System," *Neutron Radiography (3): Proceedings of the Third World Conference*, S. Fujine, K. Kanda, G. Matsumoto and J. P. Barton, Eds., Kluwer Academic Press, Dordrecht, The Netherlands, 1990, pp. 875–883.

[8] Markgraf, J. F. W., "Neutron Radiography Working Group," *Neutron Radiography: Proceedings of the Second World Conference*, J. P. Barton, G. Farny, J. Person, and H. Röttger, Eds., D. Reidel, Dordrecht, Holland, 1987, pp. 59–67.

[9] Domanus, J. C., "Can Neutron Beam Components and Radiographic Image Quality Be Determined by the Use of Beam Purity and Sensitivity Indicators?," *Neutron Radiography: Proceedings of the Second World Conference*, J. P. Barton, G. Farny, J. Person, and H. Röttger, Eds., D. Reidel, Dordrecht, Holland, 1987, pp. 839–848.

[10] Bayon, G. and Laporte, A., "Methodology Used to Characterize an Industrial Neutron Radiography Facility: Use of the Specific Image Quality Indicator," *Neutron Radiography: Proceedings of the First World Conference*, J. P. Barton and P. von der Hardt, Eds., D. Reidel, Dordrecht, Holland, 1983, pp. 1003–1011.

John D. Fenton[1]

Present[2] and Future NDT Standards for Liquid Penetrant and Magnetic Particle Testing

REFERENCE: Fenton, J. D., **"Present and Future NDT Standards for Liquid Penetrant and Magnetic Particle Testing,"** *Nondestructive Testing Standards—Present and Future, ASTM STP 1151,* H. Berger and L. Mordfin, Eds., American Society for Testing and Materials, Philadelphia, 1992, pp. 41–48.

ABSTRACT: Consider four basic types of documents—guides, practices, references, and specifications. The documents written by ASTM E07.03 on Penetrant and Magnetic Particle Methods have traditionally been guides, practices, and references. This subcommittee will now include specifications. Guides and practices instruct or guide the user, define good industry practices, and always give the user every possible option from which to select. They supply detailed *how-to* information to the user on how to apply the process. The references are a collection of graded anomalies that furnish the engineer with a very useful tool by which he can define acceptance criteria. A specification is a control document that establishes the basic parameters within which the process must be controlled. They are *thou shall* documents and supply very little how-to information.

In order to call out guides and practices on an engineering drawing, in a specification, or in a contract, many qualifying statements are required, making these documents undesirable for this purpose. Specifications supply a minimum amount of instructions or how-to information—they define the parameters within which the process is applied or controlled. There are, basically, three types of specifications: (1) those that control the process, (2) those that establish the acceptance criteria, and (3) those that do both.

KEY WORDS: nondestructive testing, fluorescent liquid penetrant testing, visible liquid penetrant testing, water washable method, post emulsified method, fluorescent magnetic particle testing, visible magnetic particle testing, continuous method, residual method

Subcommittee Structure and Current Documents

ASTM Subcommittee E07.03 is currently in the process of writing specifications for the liquid penetrant and magnetic particle methods. These documents parallel the respective military specifications and are expected to replace those documents following Department of Defense (DoD) coordination.

Responsibility for the development of standards, guides, tutorials, and reference photographs for Liquid Penetrant and Magnetic Particle Examination lies in Subcommittee E07.03. In order to properly fulfill this responsibility, the subcommittee has been divided into four sections as follows:

Section .01—Liquid Penetrant Methods, John Fenton, acting chairman
Section .02—Magnetic Particle Methods, Bernard Strauss, chairman
Section .03—Reference Photographs, Richard Gaydos, chairman
Section .04—Editorial, Calvin McKee, chairman

[1] Consultant, Durell & Associates, Fort Worth, TX 76134, and chairman of ASTM Committee E07.03.
[2] As of April 1991.

The Subcommittee currently has eleven penetrant and four magnetic particle documents either published or being balloted in ASTM and one magnetic particle document being balloted concurrently in main committee and subcommittee. These documents are as follows:

Penetrant
E 165   Practice for Liquid Penetrant Inspection Method
E 1208   Test Method for Fluorescent Liquid Penetrant Examination Using the Lipophilic Post-Emulsification Process
E 1209   Test Method for Fluorescent Penetrant Examination Using the Water-Washable Process
E 1210   Test Method for Fluorescent Penetrant Examination Using the Hydrophilic Post-Emulsification Process
E 1219   Test Method for Fluorescent Penetrant Examination Using the Solvent-Removable Process
E 1220   Test Method for Visible Penetrant Examination Using the Solvent-Removable Process
E 1418   Visible Penetrant Examination Using the Water-Washable Process
E 1135   Method for Comparing the Brightness of Fluorescent Penetrants
E 433   Reference Photographs for Liquid Penetrant Inspection
E 1316   Terminology for Nondestructive Examination (Section "F")
E 1417   Practice for Liquid Penetrant Examination
Magnetic Particle
E 709   Practice for Magnetic Particle Examination
E 1444   Practice for Magnetic Particle Examination
E 125   Reference Photographs for Magnetic Particle Indications on Ferrous Castings
E 1316   Terminology for Nondestructive Examination (Section "G")

Current Subcommittee Activity

The subcommittee has been very active the last six years, writing eleven of the above listed documents. Two of the remaining documents (E 165 and E 709) have undergone major revisions, and seven of the documents (E 1208, E 1209, E 1210, E 1219, E 1220, E 1316 Section F and E 1316 Section G) have been updated in the past two years by revision and reballoting. Two of the documents, E 1417 and E 1444, have been written to replace MIL-STD-6866 and MIL-STD-1949A, respectively.

The subcommittee initiated activity to write documents to replace military standards shortly after the government issued three documents[3] establishing DoD policy and direction to use industry consensus standards to the maximum extent possible. When the initial DoD directive was issued, there was a perception by DoD personnel that industry consensus standards would be misinterpreted or misused by being invoked across the board or used in unacceptable applications. The question was, "What control would the life cycle manager, or cognizant engineer, have over the industry document?" A process for DoD adoption of industry standards was devised by DoD to respond to these fears and concerns. The adoption process assures that the cognizant engineer retains complete control of the technical content of the document for which he is responsible or is the technical point of contact. The DoD adoption

---

[3] a. OMB Circular A-119, "Federal Participation in the Development and Use of Voluntary Standards."
  b. DoD Directive 4120.3, "Defense Standardization and Specification Program."
  c. OUSD (R&E) Publication SD-9, "DoD Interaction and Nongovernment Standardization Bodies."

procedure has been fashioned to make sure that the DoD agency that has participated in the document's development and which it will use does not lose its control over the content of, or any revision to, the document after it has been published. Details of the adoption and control procedures are in DoD 4120.3.

ASTM and SAE committees and committees of other standardizing bodies are working to convert military specifications to industry consensus standards. The decision to convert MIL-STD-6866 and MIL-STD-1949A to ASTM documents was made by Subcommittee E07.03 in Fort Lauderdale at its January 1987 meeting. The work needed to be completed in minimum time to assure that ASTM issued the first industry versions of these standards.

Current Section Activity by Section

Section .01, on Liquid Penetrant Methods, completed the first draft of the ASTM version of MIL-STD-6866 by mid-1987 and reviewed comments on the first ballot in June 1987. The first four ballots were issued with the following cover letter:

> The attached document is being considered as a direct replacement for MIL-STD-6866 on Liquid Penetrant Inspection. It has been written in the context of MIL-STD-6866 as a control document. **It is,** therefore, a document that can be called out in a specification, on an engineering drawing, or specified in a contract. By necessity, it is a **thou shall** document, not "maybe," "if you want to," "it's a good practice," etc.; it specifies the way it **will be done** unless proper authority grants a written deviation. It **is not** a tutorial such as E 165 or a detailed procedure that a floor inspector would use to apply the process.
>
> The purpose of this document is to establish the parameters within which the penetrant method will be applied, and to supply the engineer with the tools with which to specify acceptance criteria; for instance, a drawing callout may read as follows:
> *Note A:* Penetrant Inspect per ASTM E XXXX Type I, Method A or B, Sensitivity Level 3, Class 3 per Table I, Class 2 per Table II.
> This callout establishes complete control of the process and specifically defines the acceptance criteria. It requires minimum effort for the engineer. The supplier can read the drawing note and this document, and prepare a detailed procedure for application of the penetrant method to meet the intended requirements of the callout, or any other application with a similar drawing, specification, or contract callout.
>
> ASTM product committees can use this standard test method to specify acceptance criteria for products under their jurisdiction rather than writing their own documents as we have seen so many times in the past. As specifically stated in 5.6.3, this test method does not establish the acceptance criteria.
>
> Either ASTM will issue this type of document or SAE will, in which case ASTM will be relegated to writing tutorials. AMS documents will be called out when the penetrant method is to be applied.
>
> Review the document, write down your comments, and, if you would like, call me to discuss them, especially prior to a negative vote because of the nature or tone of the document.

Eight ballots were issued, the last two in both main and subcommittees. Seven negatives were received on the last main committee ballot, August 1990, and all were resolved by agreements to withdraw the negative and ballot the changes as soon as practical through sub and main committee. These changes were reviewed in Fort Lauderdale in January 1991 and submitted to ASTM for concurrent sub and main ballot in April 1991. The ASTM version of MIL-STD-6866 will be on the ASTM Society ballot in April 1991 and, if approved, will be published by midsummer. Grover Hardy, Air Force Materials Laboratory, Ohio, has reviewed and commented on all drafts and states that he "considers this document a direct replacement to MIL-STD-6866." Indications are that Mr. Hardy will coordinate the ASTM approved document in DoD to be used as a direct substitute for MIL-STD-6866.

The intent of the document is stated in paragraph 4.1 as:

4.1 This standard practice establishes the basic parameters for controlling the application of the liquid penetrant method. This standard practice is written so it can be specified on the engineering drawing, specification, or contract. It is not a detailed how-to procedure to be used by the inspector and, therefore, must be supplemented by a detailed procedure that conforms to the requirements of this standard practice. E 165 contains information to help develop detailed how to requirements.

Some of the changes in this document include the use of the daily system check to evaluate all parameters of the process. This includes those parameters currently requiring MIL-I-25135 tests. The MIL-I-25135 tests will be used, when necessary, to determine the cause of failure in the daily systems check. The paragraph reads as follows:

7.8.4 *System Checks*—The test specified in 7.8.4.1–7.8.4.4 shall be made at intervals specified in Table III. These periodic checks of penetrant materials may be waived if the known defect standard(s) selected for the system performance check adequately monitors the serviceability of the penetrant materials and the results of the daily performance checks are documented in sufficient detail to allow an audit to detect deterioration of performance below satisfactory levels.
  7.8.4.1 *Penetrant Brightness*—
  7.8.4.2 *Penetrant Removability (Method A only)*—
  7.8.4.3 *Penetrant Sensitivity*—
  7.8.4.4 *Emulsifier Removability*—

Incorporated in this document are tables that permit the engineer to establish acceptance criteria. This document does not establish acceptance criteria as noted in paragraph 6.2:

6.2 *Specifying*—When examination is required in accordance with this standard, orders, contracts or other appropriate documents, they shall indicate the criteria by which components are judged acceptable. An example of such criteria is contained in Tables I and II, and in MIL-STD-1907; however, other criteria may be utilized. Engineering drawings or other applicable documents shall indicate the acceptance criteria for the entire component; zoning may be used. Inspection on a sampling basis shall not be allowed unless specifically permitted by the contract.

The engineer can use the tables included, or other sources, to establish accept/reject criteria. Table I establishes gradations for rounded indications and Table II establishes gradations for elongated indications.
  Personnel qualification and certification will be to ASNT SNT-TC-1A for commercial work and MIL-STD-410 for military work as specified by the purchase order or contract. Restrictions and usage within those documents will be the controlling requirements.

6.3 *Personnel Qualification*—Personnel performing examinations to this document shall be qualified and certified in accordance with SNT-TC-1A or MIL-STD-410 for military purposes, or as specified in the contract or purchase order.

E 165 has been completely revised to be used as a guide or tutorial to support E 1417. It serves as a guide when selecting type, method, and sensitivity, and for general information about the penetrant method.
  Separate documents (E 1208, E 1209, E 1210, E 1219, E 1220, and E 1418) have been written as standard test methods and cover specific penetrant types and methods. They were designed to be used (1) to ascertain the applicability and completeness of a company's process, (2) in the preparation of process specifications, and (3) in the organization of facilities.

TABLE I—*Maximum permissible rounded indications.*[1]

Class	Indication Maximum Dimension, in inches	Less Than 0.040	0.040–0.069	0.070–0.104	0.105–0.154	0.155–0.229	0.230–0.339	0.340–0.489	0.490 and Greater
				Material Thickness Range, In.					
1	Any length	0	0	0	0	0	0	0	0
2	Up to 1/64	0	0	1	2	2	3	3	4
3	Up to 1/64	0	1	2	2	3	4	4	5
	>1/64–2/64	0	0	0	1	2	3	3	4
4	Up to 1/64	1	2	2	3	3	4	5	6
	>1/64–2/64	0	0	1	2	2	3	4	5
	>2/64–3/64	0	0	0	1	1	2	3	4
5	Up to 1/64	2	3	3	4	4	5	6	7
	>1/64–2/64	1	2	2	3	3	4	5	6
	>2/64–3/64	0	1	1	2	3	3	4	5
	>3/64–4/64	0	0	0	1	1	2	3	4
6	Specific acceptance limits must be agreed upon between purchaser and supplier.								

Note: Metric conversion: The dimensions are in English units. Multiply English units by 25.4 to convert to millimetres.

[1] The maximum number of permissible indications relates to 6 in.^{2} of surface area with the major dimension of the containment area not to exceed 6 in.

TABLE II—*Maximum permissible linear indications.*[1,2]

Class	Indication Length (in inches)	Less Than 0.040	0.040–0.069	0.070–0.104	0.105–0.154	0.155–0.229	0.230–0.339	0.340–0.489	0.490 and Greater
				Material Thickness Range, In.					
1	Any length	0	0	0	0	0	0	0	0
2	Up to 2/64	0	0	0	0	1	2	2	3
3	Up to 2/64	0	0	1	2	3	4	4	5
	>2/64–4/64	0	0	0	1	2	3	4	4
4	Up to 2/64	0	1	2	3	4	4	5	5
	>2/64–4/64	0	0	0	1	2	3	4	4
	>4/64–6/64	0	0	0	0	0	0	1	2
5	Up to 2/64	1	2	2	4	4	5	6	6
	>2/64–4/64	0	1	2	2	4	4	5	5
	>4/64–8/64	0	0	0	1	2	3	4	4
	>8/64–16/64	0	0	0	0	0	1	2	3
6	Specific acceptance limits must be agreed upon between purchaser and supplier.								

Note: Metric conversion: The dimensions are in English units. Multiply English units by 25.4 to convert to millimetres.

[1] The maximum number of permissible indications relates to 6 in.^{2} of surface area with the major dimension of the containment area not to exceed 6 in.

[2] Any indication must be separated from any other indication (edge to edge) by a minimum distance equal to the maximum length of the largest of the two indications.

Section .02, on Magnetic Particle Method, started the rewrite of MIL-STD-1949A about the same time as Section .03 started the ASTM version of MIL-STD-6866. Since a major rewrite of MIL-STD-1949A was under way by the Army, a decision was made to delay the ASTM version until several issues raised during DoD coordination of the new draft, MIL-STD-1949B, could be resolved. The ASTM version of MIL-STD-1949 was balloted in sub and main committees in August 1990. One negative was received and has been resolved. Changes were approved in subcommittee in January 1991 and the document was approved for reballot in April 1991.

Paragraph 1.1 states the intent of this document:

> 1.1 This standard practice establishes minimum requirements for magnetic particle examination used for detection of surface or slightly subsurface discontinuities in ferromagnetic material. This document is intended as a direct replacement of MIL-STD-1949. ASTM 709 can be used in conjunction with this document as a tutorial.

This document is considered by DoD as a direct replacement to MIL-STD-1949B (DoD coordinated but not issued). The ASTM version is, for all practical purposes, identical to MIL-STD-1949B, which has passed DoD coordination. With acceptance on the April ASTM society ballot, this document could be published by ASTM by late summer and through DoD coordination by early 1992.

E 709 has undergone major revisions to bring it up to date and expand its contents. This document, like E 165, has been written as a guide or tutorial to support the new "thou shalt" document, the ASTM version of MIL-STD-1949. Three negatives were received on the last ballot and have been resolved. This document, E 709, was revised, reviewed in Fort Lauderdale in January 1991, and approved for reballot. It is being balloted concurrently in sub and main committees on the April 1991 ballot. Hopefully, the new E 709 will be issued in 1991.

Section .02 is preparing two documents on thin metal strips containing artificial defects, shims, to be used when setting up magnetic particle techniques. They are (1) a standard for the control and fabrication of shims and (2) a procedure to control round robin tests to determine the practical use and limitations of shims. The section plans a document similar to E 1025 on Hole Type Image Quality Indicators Used in Radiography. It is anticipated that this document would control the material, slot size, and similar characteristics of the shims. The round robin would use either existing, or specially fabricated shims, or both, in an attempt to determine the limitations within which the shims are practical to use. Consideration will be given to material type, condition, configuration, and similar characteristics. Applicable portions of this information will be incorporated into E 709 and the ASTM version of MIL-STD-1949.

Section .03, on Reference Photographs, has made numerous requests for support to develop reference photographs for both penetrant and magnetic particle examination to replace E 125 and E 433 with new documents. Some photographs have been received and would be acceptable if additional photographs could be obtained to fill the gaps. Boeing Airplane Co. is in the process of collecting sample material and parts with natural occurring defects to develop a set of penetrant reference photographs. It appears at this time that Boeing will be willing to supply ASTM with a set of the photographs for publication. Boeing's data will be pertinent information and will cover material type and condition, metallographic data on the defect, and photographs of the indications. The intent is to develop a set of reference plates similar to the reference radiographs. Subcommittee members have been requested to support this effort by sending samples of defective material and parts to Boeing. Reference photographs for magnetic particle examination still remains a problem. Good reproducible photographs of indications with all of the pertinent information [material type, condition, magnetizing method, actual indication size (not defect length), photographic technique] are needed. Sample material will be accepted as long as the samples can be easily handled and can be cut.

Section .04, Editorial, has just completed merging the two subcommittee documents, E 269 and E 270, into E 1316 and adding several new terms. E 1316 is a collection of all terms defined and used by all subcommittees of Committee E-7. E 1316 contains a general section and ten alphabetized sections. The general section includes terms used by all subcommittees; Sections F and G contain terms used in magnetic particle and penetrant testing, respectively. E 269 and E 270 have been cancelled.

Future activities for Subcommittee E07.03 include:

1. Continue pursuing additional avenues to develop a set of graded references, possibly sketches rather than photos, for both penetrant and magnetic particle.
2. Incorporate controls for automated penetrant systems into E 1417.
3. Develop a standard for the fabrication of shims.
4. Conduct a round robin on the application of shims.

The subcommittee plans to develop standards so they fit the following format:

Penetrant
  Guide or Tutorial
  E 165   Practice for Liquid Penetrant Examination
  Standard Practice
  E 1417   Practice for Liquid Penetrant Examination
  Reference Photographs
  E 433   Reference Photographs for Liquid Penetrant Inspection
  E XXXX   Graded Reference Photographs
  Standard Method
  E 1208   Test Method for Fluorescent Liquid Penetrant Examination Using the Lipophilic Post-Emulsification Process
  E 1209   Test Method for Fluorescent Penetrant Examination Using the Water-Washable Process
  E 1210   Test Method for Fluorescent Penetrant Examination Using the Hydrophilic Post-Emulsification Process
  E 1219   Test Method for Fluorescent Penetrant Examination Using the Solvent-Removable Process
  E 1220   Test Method for Visible Penetrant Examination Using the Solvent-Removable Process
  E 1418   Test Method for Visible Penetrant Examination Using the Water-Washable Process
  E 1135   Method for Comparing the Brightness of Fluorescent Penetrants
  Terminology
  E 1316   Terminology for Nondestructive Evaluations (Section "F")
Magnetic Particle
  Guide or Tutorial
  E 709   Practice for Magnetic Particle Examination
  Standard Practice
  E 1444   Standard Practice for Magnetic Particle Examination

Reference Photographs
  E 125   Reference Photographs for Magnetic Particle Indications on Ferrous
              Castings
Standard Method
  E XXXX   To be decided in subcommittee
Terminology
  E 1316   Terminology for Nondestructive Evaluations (Section "G")

In summary, the subcommittee has been very active. We have accomplished a lot of work and have a lot to do; therefore, we will continue to be very active and would appreciate the participation of anyone that wants to join the Subcommittee.

Howard E. VanValkenburg[1]

Standardization of Ultrasonic NDT from an ASTM Perspective: Past, Present, and Future

REFERENCE: VanValkenburg, H. E., **"Standardization of Ultrasonic NDT from an ASTM Perspective: Past, Present, and Future,"** *Nondestructive Testing Standards—Present and Future, ASTM STP 1151,* H. Berger and L. Mordfin, Eds., American Society for Testing and Materials, Philadelphia, 1992, pp. 49–55.

ABSTRACT: ASTM ultrasonic NDT standards developed by Subcommittee E07.06 range in scope from tutorial documents to definitive procedures. The evolution of these has paralleled that of the underlying technology, which is now approaching 50 years of practical application. The chronology of this relationship and its importance in promoting acceptance of the methodology are discussed. The scope and status of current standards, in-process revisions, developmental documents, and recommended future projects are reviewed. Some of the many other sources for standards-related documents, both within and outside of ASTM, are summarized.

KEY WORDS: nondestructive testing, ultrasonic nondestructive testing, technology, standards, chronology, Subcommittee E07.06

The Past

Origins of the Technology

While both theoretical studies and laboratory investigations of physical acoustics at ultrasonic frequencies began in the 19th century, proposals to apply that science to inspection of materials first appeared in the late 1930s. Various techniques utilizing continuous waves (c-w) and what is now called "through transmission" appear to have been the only approaches tried. Source references can be found in Bergmann [1], Hastings and Carter [2], McMaster [3], Graff [4], and McIntire [5]. Limited success was reported in the detection of gross flaws such as laminations in metal sheet. Calibration or standardization would hardly have been a concern, although artificial defects must surely have been used to validate these early tests.

The technology changed abruptly as the result of two patents granted to American inventors. In 1942 one for a flaw detection and measurement method using pulsed ultrasonic wave trains was issued to Floyd A. Firestone [6]. Within two years commercialization of his Supersonic Reflectoscope was under way, and many significant industrial applications were being reported [7]. Some of the earliest publications describe the use of drilled holes to prove test capability and to establish sensitivity. In 1947 Wesley S. Erwin received a patent for an instrument called a Sonigage, which used an ultrasonic resonance technique to measure thickness [8]. Several commercial versions were already being produced, and many applications, including detection of unbonds, were found. The two methods were actually complementary, and each opened new vistas for NDT practitioners. With access to only one surface, solid objects could now be inspected for internal defects or measured for thickness. This included the detection of many discontinuity types and determination of material conditions not possible with any other existing NDT procedures.

[1] Senior research engineer, retired, Box 37, Candlewood Isle, New Fairfield, CT 06812.

For many engineering materials, part length or flaw size did not appear to be the principal limiting factors. Rather it was the great range of test capability, sensitivity to even minute discontinuities such as grain boundaries, dependence on instrumentation settings, and interpretation of test results that were perceived as formidable obstacles. For example, there was no direct equivalent of the radiographic film image and the trusted penetrameter. It was soon apparent to many that full utilization of these test methods would require a multifold approach involving: (1) education, (2) demonstration, and (3) standardization.

A Chronology of the Standardization Process

While several of the established technical societies were involved almost from the start of industrial ultrasonic testing, emphasis in this review is on the contributions of ASTM Committee E07. Originally concerned only with radiography, its scope had expanded in 1947 to include other NDT methods. It should be mentioned that concurrently the fledgling organization of specialists, which had begun in 1940 as the American X-Ray and Radium Society, had also seen the future, changed its name to the Society for Nondestructive Testing, and renamed its journal accordingly. As is still the case today, many individuals who contributed to the development of NDT were active in both groups. In ASTM the principal thrust was, of course, on standards, while in ASNT it was on applications. As a precursor to standardization efforts, ASTM Committee E07 sponsored several valuable educational symposiums, leading in 1951 to the landmark technical publication *Symposium on Ultrasonic Testing, STP 101* [2]. The contents comprised presentations made at three different meetings:

1. 1948 Annual Meeting—Round Table Discussion on Ultrasonic Testing

   Louis Gold, "Ultrasonic Wave Propagation in Materials"
   C. H. Hastings and S. W. Carter, "Inspection, Processing and Manufacturing Control of Metals by Ultrasonic Methods"

2. 1949 Annual Meeting—Symposium on Ultrasonic Testing

   J. C. Smack, "Basic Principles of Ultrasonic Testing"
   J. C. Hartley, "Ultrasonics in the Heavy Forging Industry"
   J. V. Carroll, "The Application of Ultrasonics to the Fabrication of Aluminum"
   E. D. Hall, "Ultrasonic Testing in Railroad Work"
   D. M. Kelman, "Ultrasonics in the Electrical Industry"
   A. Piltch, "Ultrasonic Testing of Bronze Forgings and Ingots"

3. 1950 Annual Meeting—Technical Papers

   J. R. Leslie, "Pulse Techniques Applied to Dynamic Testing"
   A. G. H. Dietz et al., "The Measurement of Dynamic Modulus in Adhesive Joints at Ultrasonic Frequencies"

It is evident that ultrasonic methods were by then in widespread use; however, such topics as calibration, standardization, specifications, and acceptance criteria are hardly mentioned. Smack does illustrate two "test blocks" used for setting up a reflectoscope and to provide a "standard" for uniform testing from day to day. He states: "So far no attempt has been made to establish industry-wide standards. Individual manufacturers have set their own standards to suit their own products and customers' requirements." Other authors present similar viewpoints. In 1950, Committee E07 organized Subcommittee VI on Ultrasonic Testing Methods, with John Smack as chairman.

The first two standards developed by the subcommittee were basically tutorial:

E 113-55T—Tentative Recommended Practice for Ultrasonic Testing by the Resonance
Method
E 114-55T—Tentative Recommended Practice for Ultrasonic Testing by the Pulsed Contact Method

The chairman's 1956 annual report to E07 states that there were then in process five proposed new standards relating to these subjects:

Terminology
Angle-Beam Test Methods
Inspection of Butt Welds
Standardization of Reference Blocks
Recommendations for Inspector Qualification

Each of these, except the last, did eventually proceed through the consensus process and issue as a standard under the jurisdiction of E07. The last was undoubtedly one of the Subcommittee's most short-lived projects, as the task group formed to draft a document was dissolved almost immediately when the controversial implications of the project were recognized. Many years later ASNT did successfully confront the personnel qualification issue.

Related ASTM Contributions to Early NDT Acceptance

During these formative years ASTM, either singly or as a cosponsor, provided a forum through meetings, lectures, and publications for the promotion of research, development, and application in all NDT methods. Listed below are some which resulted in STPs, each containing useful references to contemporary ultrasonic NDT technology.

1950—*Role of NDT in the Economics of Production*	(STP 112)
1952—*Determination of Elastic Constants*	(STP 129)
1952—*Nondestructive Testing (held in New York)*	(STP 145)
1956—*Nondestructive Testing (held in Los Angeles)*	(STP 213)
1957—*NDT in the Field of Nuclear Energy*	(STP 223)
1959—*NDT in the Missile Industry*	(STP 278)

Also in 1952, Robert C. McMaster, invited to present the Edgar Marburg Honor Lecture, not surprisingly chose as his topic "Nondestructive Testing" [9].

During the ensuing years, several additional ultrasonic test standards were developed by the Subcommittee; one was acquired (from E04), and one was withdrawn (E 113). At the end of the past period, that is, April 1991, E07 had formal responsibility for 20 issued standards. Finally, in summarizing this period, it should be noted that in 1976 ASTM cosponsored with NIST (then NBS) and ASNT the first Symposium on Nondestructive Test Standards and published the proceedings as STP 624 [10]. Subcommittee members contributed the following three key papers on ultrasonics:

J. E. Bobbin, "Ultrasonic Standards—Overview"
J. T. McElroy, "Search Unit Standardization"
C. E. Burley, "Calibration Blocks for Ultrasonic Testing"

The Present

Review of Current E07.06 Standards

The 20 ASTM standards for which the Subcommittee has responsibility fall within three basic categories: terminology, practices, and guides. All appear in Volume 03.03 of the Annual Book of ASTM Standards [11] and consist of the following:

Glossary-Ultrasonic NDT—Part I of Standard Terminology E 1316
Basic methodologies
    Contact testing—Practice E 114
    Immersion testing—Practice E 214
    Detection and evaluation by immersion tests—Practice E 1001
    Contact angle-beam methods—Practice E 587
Inspection of specific product geometries, fabrication or type defect
    Pipe and tubing—Practice E 213
    Longitudinally welded pipe—Practice E 273
    Steel with convex, cylindrical entry surface—Practice E 1315
    Weldments by contact methods—Practice E 164
    Macro inclusions in bearing quality steel—Practice E 588
Determination of physical properties of materials
    Ultrasonic velocity—Practice E 494
    Apparent attenuation—Practice E 664
Dimensional measurement
    Thickness by the pulse-echo contact method—Practice E 797
Evaluation of inspection apparatus
    Field checking of test systems—Practice E 317
    Laboratory characterization of search units—Guide E 1065
    Electronic evaluation of instrumentation—Guide E 1324
Fabrication, control and standardization of reference blocks
    Aluminum alloy standard reference blocks—Practice E 127
    Steel and other metal reference blocks—Practice E 428
    Test blocks from production material—Guide E 1158
    Calibration by extrapolation between FBH sizes—Practice E 804

Standards Considered for Major Revisions

All ASTM standards must be reviewed within a mandatory five-year interval to determine whether they are to be revised, reapproved, or withdrawn. While relatively minor changes are often made, major ones which significantly alter the technical content are undertaken only when there is clearly justification for the effort required and the possible impact on users. These are most often necessitated by advances in the relevant technology.

Practice E 127 (aluminum standard reference blocks) is now being revised to (1) replace existing requirements for the use of obsolete instrumentation and search units, and (2) provide for calibration of inspection systems at higher test sensitivity, improved near-surface resolution, and longer metal travel. Since experimental confirmation by round-robin tests is involved, such a program requires considerable task group input and many months to complete.

Practice E 428 (steel and other alloy test blocks) is being considered for major changes to (1) limit the scope to the type of blocks made from standard bar stock, which are representative of sets supplied commercially, and (2) delete parts now better covered by the new Guide E 1158.

The subcommittee presently has no requests on record for extensive revision of any other current standards.

Three New Standards in Process

As part of the on-going E07 program to provide ASTM consensus standards to replace related military NDT documents, the Subcommittee serves as the forum to review and ballot the proposed reformatting of MIL-STD-2154, "Inspection, Ultrasonic, Wrought Metals, Process For." A second project involves development of standard guides for detection and sizing of surface-breaking discontinuities such as those due to intergranular stress-corrosion cracking. The third document[2] is titled "Guide for Data Fields for Computerized Transfer of Digital Ultrasonic Testing Data."

Two or more drafts of the proposed standards have already been balloted at subcommittee level and are now being further revised to accommodate persuasive negative votes and constructive comments.

Commentary on Some Specific E07.06 Contributions

Many of the Subcommittee's standards are referenced or specified by a diversity of users—commercial, military, and other technical societies. Basic methodology documents such as E 114 (contact testing) and E 214 (immersion testing) have been extensively used historically for tutorial purposes. Sections of others, such as E 317 (field checking of test equipment), are frequently cited for specific test requirements, vertical linearity, for example. Practice E 164 (inspection of weldments), in addition to its principal function, is the source for consensus terminology on so-called "IIW Type" calibration blocks, and a drawing for an approved version is dimensioned in U.S. customary units.

Since its original issue in 1958, Practice E 127 governing aluminum alloy reference blocks has been the only national standard which prescribes not only the material and dimensional requirements for such blocks, but also establishes definitive criteria for their ultrasonic response.

More recently the subcommittee has developed two new standards for which there has been a long-standing need. Both involve characterization of ultrasonic test system components by measurement of parameters which affect final performance. Guide E 1065 relates to search units and Guide E 1324 to the electronic instrumentation.

Other Sources of Ultrasonic NDT Standards Documents

In reviewing the present status of ultrasonic standards it must be noted that, as with all NDT methodologies, there are innumerable related documents originating from sources other than Committee E07. For example, a recent survey of all ASTM standards [12] identified 77 involving ultrasonic testing which came from other committees. Table 1 summarizes these with respect to the general committee groups which developed them and categories of the standards. Of all standards-related NDT documents issued, by far the greatest percentage have originated outside ASTM, although many cite its standards as general references or for specific requirements. The sources may be broadly classed as industry, government, trade organizations, and technical societies. ASNT publishes a comprehensive listing of the most widely used

[2] Approved as ASTM Standard E 1454 April 1992.

TABLE 1

Committee Group	Number Issued	Category
Metals, ferrous	56	Specifications
	1	Practice
Metals, nonferrous	14	Specifications
	1	Method
Concrete	1	Test method
Rock	1	Test method
Nuclear	1	Practice
Polymers	1	Practice
Corrosion	1	Practice

specifications and standards [13]. Since a number of other presentations in this symposium include either an overview of this subject or a detailed review of specific standards, only a few particularly relevant to ultrasonics are briefly mentioned here.

As early users of ultrasonic inspection, many industries required specifications for both test procedures and acceptance criteria. Major contributors have been the electrical machinery, chemical, petroleum, basic metals, and aerospace sectors. Examples of representative specifications include (1) General Electric P3TF1 and Pratt and Whitney SIM-1 for engine rotor forgings, and (2) McDonnell Douglas PS 21211 series for airframe components.

Federal government NDT documents originate principally from groups within DoD and DOT, although it is now official policy to encourage use of nongovernment consensus processes. As stated earlier, reformatting of several military documents such as MIL-STD 2154 has been undertaken by subcommittees of E07.

Among the technical societies and trade organizations, the major originators of NDT standards involving ultrasonics include the Aluminum Association, API, ASME, AWS, and SAE. Parallel standards for medical diagnostic instrumentation are being developed by the American Institute for Ultrasound in Medicine (AIUM).

The Future—A Role for Subcommittee E07.06

The Broad Objectives

The most often heard challenge to the NDT profession is for improved quantification of the materials evaluation process. Ultrasonic methods in particular have been identified as having present shortcomings but great potential. Among the perceived limitations are the reproducibility and precision of test results. In the standardization process more effective use must be made of interlaboratory studies and the confirming documentation made permanently available as ASTM research reports. Terminology standards should be revised frequently to reflect new or changed technology and thus avoid the confusion that can result from ambiguous, conflicting, misleading, or undefined terms. Practices for characterizing performance-related parameters of test systems must be current, definitive, and encompass all components. New methods and apparatus should be reviewed as they are proved practical to determine if existing standards can be applied or new ones need to be developed.

Specific Recommendations

While the following are the author's personal recommendations for future Subcommittee action, they are believed to be consistent with those discussed and approved at past meetings of the sections and task groups involved.

1. Complete the in-process and proposed revisions of standards relating to reference blocks, contact method, angle beam calibration, search units, and instrumentation.
2. Amend practices for measurement of attenuation, thickness, and velocity to provide more detailed reproducibility and precision statements with substantiating documentation from round-robin studies.
3. Place more emphasis on limitations, precautions, and test anomalies where appropriate.
4. Continue conversion of parallel military standards to ASTM format through the E07 consensus process.
5. Provide more thorough review of proposed ISO/TC-135 ultrasonic test documents in an effort to achieve consistency with E07.06 standards, avoid duplication of input, and fully utilize the technical expertise available.
6. Intensify efforts to achieve improved liaison and cooperation with other ASTM committees, particularly those involved with product specifications.
7. Pursue standardization of newer test methods such as those utilizing ultrasonic imaging, signal processing, transducer arrays, and computer interfacing.

References

[1] Bergmann, L., *Der Ultraschall,* Hirzel-Verlag, Stuttgart, 1954.
[2] Hastings, C. E. and Carter, S. W., "Inspection, Processing and Manufacturing Control of Metals by Ultrasonic Methods," *Symposium on Ultrasonic Testing, STP 101,* American Society for Testing and Materials, Philadelphia, 1951.
[3] *Nondestructive Testing Handbook,* R. C. McMaster, Ed., Sects. 43–46, Ronald Press, New York, 1959.
[4] Graff, K. F., "Ultrasonics: Historical Aspects," *IEEE Symposium on Sonics and Ultrasonics—1977,* 77CH1264-ISU, Institute of Electrical and Electronics Engineers, New York, 1977.
[5] *Nondestructive Testing Handbook,* 2nd ed., Vol. 7, *Ultrasonic Testing,* P. McIntire, Ed., American Society for Nondestructive Testing, Columbus, OH, 1991.
[6] Firestone, F. A., U.S. Patent 2,280,226, 21 April 1942.
[7] Carlin, B., "Supersonic Examination of Materials," *Product Engineering,* October 1947.
[8] Erwin, W. S., U.S. Patent 2,431,233, 18 Nov. 1947.
[9] McMaster, R. C., "Nondestructive Testing," Marburg Lecture, *Annual Proceedings,* American Society for Testing and Materials, Philadelphia, 1952.
[10] *Nondestructive Testing Standards—A Review, STP 624,* H. Berger, Ed., American Society for Testing and Materials, Philadelphia, 1977.
[11] *Annual Book of ASTM Standards,* Vol. 03.03 (Nondestructive Testing), American Society for Testing and Materials, Philadelphia, 1990.
[12] Standards Related to Nondestructive Testing Developed by ASTM Committees Other Than Committee E-7, *Annual Book of ASTM Standards,* Vol. 03.03 (Nondestructive Testing), American Society for Testing and Materials, Philadelphia, 1988, pp. 629–630.
[13] *NDT Specifications and Standards Commonly Used; List of Issuing Organizations,* Pub. No. 113, American Society for Nondestructive Testing, Columbus, OH, 1988.

William D. Jolly[1]

Status and Future Directions for Acoustic Emission Standards

REFERENCE: Jolly, W. D., "Status and Future Directions for Acoustic Emission Standards," *Nondestructive Testing Standards—Present and Future, ASTM STP 1151,* H. Berger and L. Mordfin, Eds., American Society for Testing and Materials, Philadelphia, 1992, pp. 56–62.

ABSTRACT: Acoustic emission is still a dynamic technology after some 25 years of continuing development. Standards are developed as new applications are proven, and the need for new standards is pressed by both suppliers and users of acoustic emission services. At the present time, ASTM has twelve acoustic emission standards in Volume 03.03 of the *Annual Book of ASTM Standards* and five in development; the American Society of Mechanical Engineers Section V has two acoustic emission standards; the Society for the Plastics Industry has one; and the American Society for Nondestructive Testing now offers testing for certification of acoustic emission practitioners.

Acoustic emission standards are typically written by representatives of the users, suppliers of services and instruments, and researchers. This mix assures that standards will satisfy the needs of the community and incorporate new technology. Some acoustic emission standards anticipated for the future are offshore testing, underwater surveillance, and aircraft structural-integrity assessment.

KEY WORDS: acoustic emission, AE standards, acousto/ultrasound

Acoustic Emission Technology

Acoustic emission (AE) is defined in ASTM Terminology for Nondestructive Evaluations (E 1316) as "The class of phenomena whereby transient elastic waves are generated by the rapid release of energy from localized sources within a material, or the transient elastic waves so generated." A localized failure releases transient elastic waves, which on the planetary scale are called earthquakes. In smaller structures such as pressure vessels or bridges, the term AE was coined because in early examples the transients were audible. One example is called "Tin Cry."

In the late 1960s AE was applied as an effective means of assessing the integrity of structures and pressure vessels [1]. In 1967 Jack Spanner, Sr., coordinated the formation of the Acoustic Emission Working Group, and AE instrumentation components such as sensors and preamplifiers became commercially available [2]. Research led to more diverse applications, and the need for standards became apparent. ASTM Subcommittee E7.04 was organized by Spanner in June 1972 to promote the advancement of AE technology and the formulation of AE standards and included researchers, vendors, and users. By 1975 the use of AE as a nondestructive testing tool was widely accepted by industry [3]. The first ASTM standard on AE approved by ballot in 1976 was designated E 569–76: Recommended Practice for Acoustic Emission Monitoring of Structures During Controlled Stimulation.

[1] Institute scientist, Southwest Research Institute, 6220 Culebra Rd., San Antonio, TX 78228-0510.

Application Areas

AE has application where the stresses on a structure stimulate the release of energy as a detectable indication of the possible future failure of the structure. AE is used to detect and locate cracks in hydroelectric dams, assess the integrity of highway bridges, qualify fiberglass tanks and vessels for service, inspect fiberglass bucket truck booms, monitor the quality of spot welds, and requalify pressure vessels and piping in the petrochemical industry. AE is commonly used to monitor the progress of fatigue testing on many types of structures, and AE instrumentation is installed on critical spars in the wing of some C5A aircraft. These examples illustrate the range of AE applications but not the extent to which AE is used today.

The instrumentation and techniques for AE monitoring are also applied to detect leaks in pressurized systems and to assess damage in bearings and, in the case of acousto/ultrasound, to detect defective parts by evaluation of response of the part to insonification.

Functional Aspects

AE is useful as a nondestructive evaluation (NDE) method when controlled stimulation causes the release of transient acoustic energy bursts from defects in a structure. Piezoelectric sensors mounted on the structure convert the acoustic signals to electrical waves which are amplified, counted, integrated, timed, or otherwise processed to produce a record of the AE response to the applied stimulus. Interpretation of this response leads to the detection of defects in the structure or the qualification of the structure for its intended service.

Pioneers in AE testing developed an experience base from which to judge the condition of a structure established on the observed AE response. As this lore was difficult to pass on to service technicians except by hands-on experience, it became necessary to formalize the procedures for calibration, operation, and interpretation of data from AE instruments. Because each practitioner and vendor developed their own procedures, they were difficult to compare because of instrument differences and the lack of a common sensor-sensitivity calibration method.

Standards Requirement

In order to establish a common basis for comparing the results obtained from different instruments, it was necessary to develop standards for the conduct of AE testing. Instruments from different vendors offered a variety of terms for very similar output quantities. An example of this was the measured area under the rectified signal envelope. This quantity has been called energy, marse, energy release, signal strength, and relative AE signal strength, to name a few. These measures of AE signals were useful in a general sense, but the quantity had no engineering definition. To this day, the quantity persists, however, while the AE community still struggles with the development of a meaningful definition.

Another problem area is the development of a standard for secondary calibration of AE sensors. A primary AE sensor calibration standard was developed at the National Institute of Standards and Technology (NIST), but vendors of AE services need an affordable laboratory procedure that can be performed frequently to verify the continued compliance of sensors. The primary calibration procedure requires prohibitively large and expensive equipment for the small AE service vendor. A secondary calibration standard must await the completion of a secondary calibration procedure that is being developed at NIST.

Existing Standards

Methods

ASTM E 1106: Test Method for Primary Calibration of Acoustic Emission Sensors—This method covers the requirements for the absolute calibration of AE sensors. The calibration yields the frequency response of a transducer to waves at a surface of the type normally encountered in AE work.

Practices

ASTM E 1118–89: Practice for Acoustic Emission Examination of Reinforced Thermosetting Resin Pipe—This practice covers AE examination or monitoring of reinforced thermosetting resin pipe to determine structural integrity. It is applicable to lined or unlined pipe, piping systems, fittings, and joints.

ASTM E 1139–87: Practice for Continuous Monitoring of Acoustic Emission from Metal Pressure Boundaries—This practice provides guidelines for continuous monitoring of AE from metal pressure boundaries in industrial systems during operation. Examples are piping, pressure vessels, and other system components which serve to contain system pressure. Pressure boundaries other than metal such as composites are specifically not covered by this document.

ASTM E 749–80 (Reapproved 1985): Practice for Acoustic Emission Monitoring During Continuous Welding—This practice provides recommended guidelines for AE monitoring of weldments during and immediately following their fabrication by continuous welding processes. The technique is in a developmental stage and is not used routinely on production welding.

ASTM E 569–85: Recommended Practice for Acoustic Emission Monitoring of Structures During Controlled Stimulation—This practice provides guidelines for AE examination or monitoring of structures, such as pressure vessels, and piping systems, that can be stressed by mechanical or thermal means.

ASTM E 751–80 (1985): Practice for Acoustic Emission Monitoring During Resistance Spot-Welding—This practice describes procedures for the measurement, processing, and interpretation of the AE response associated with selected stages of the resistance spot-welding process.

ASTM E 1067–85: Practice for Acoustic Emission Testing of Fiberglass Reinforced Plastic Resin (FRP) Tanks/Vessels—This practice covers AE examination or monitoring of fiberglass-reinforced plastic tanks and vessels (equipment) under pressure or vacuum to determine structural integrity.

ASTM E 750–88: Practice for Characterizing Acoustic Emission Instrumentation—This practice is recommended for use in testing and measuring operating characteristics of AE electronic components or units. It is not intended to be used for routine checks of AE instrumentation, but rather for periodic calibration or in the event of a malfunction.

ASTM E 1211–87: Practice for Leak Detection and Location Using Surface-Mounted Acoustic Emission Sensors—This practice describes a passive method for detecting and locating the steady-state source of gas and liquid leaking from a pressurized system. The method employs surface-mounted AE sensors or sensors attached to the system via acoustic waveguides and may be used for continuous in-service and hydrotest monitoring of piping and pressure-vessel systems.

Terminology

ASTM 610–89: Standard Terminology Relating to Acoustic Emission—This standard delineates the accepted terminology used throughout industry when discussing AE technology.

Guides

ASTM E 976–84 (1988): Guide for Determining the Reproducibility of Acoustic Emission Sensor Response—This guide defines simple economical procedures for testing or comparing the performance of AE sensors. These procedures allow the user to check for degradation of a sensor or to select sets of sensors with nearly identical performances. The procedures are not capable of providing an absolute calibration of the sensor, nor do they assure transferability of data sets among organizations.

ASTM E 650–85: Guide for Mounting Piezoelectric Acoustic Emission Sensors—This document provides guidelines for mounting piezoelectric AE sensors.

New Standards

A new standard, ASTM E-1419: Test Method for Examination of Seamless, Gas-filled, Pressure Vessels Using Acoustic Emission, is currently completing the consensus process. This standard was developed from a Department of Transportation (DOT) procedure for requalification of compressed-gas transport cylinders. It will provide guidelines for AE tests of seamless pressure vessels (tubes) of the type used for distribution of industrial gases.

Other Standards

The Society of Plastics Industries (SPI) and the American Society of Mechanical Engineers (ASME) have developed standards for the inspection of fiber-reinforced tanks and vessels, and ASME has developed a standard governing the use of AE to test new pressure vessels. These are described as follows.

SPI, Reinforced Plastics and Composites Institute: Recommended Practice for Acoustic Emission Testing of Fiberglass Tanks/Vessels—This recommended practice provides guidelines for AE examination or monitoring of fiber-reinforced plastic (FRP) tanks and vessels (equipment) under pressure or vacuum to determine structural integrity.

ASME Code, Section V, Article 11: Acoustic Emission Examination of Fiber-Reinforced Plastic Vessels—This article describes or references requirements for applying AE examination of new and in-service FRP vessels under pressure, vacuum, or other applied stress.

ASME Code, Section V, Article 12: Acoustic Emission Examination of Metallic Vessels During Pressure Testing—This article describes methods for conducting AE examination of new metallic pressure vessels during acceptance pressure testing when specified by a referencing ASME code section.

Emerging Standards

New standards are generated as the technology broadens and new techniques are developed. One example is the methodology lumped under the name of acousto/ultrasound (A/U). The name derives from the early experimental use of AE sensors to detect the low-frequency response of a specimen to the impulsive excitation provided by an ultrasonic transducer. A term more descriptive of the phenomena would be mechanical spectroscopy.

The A/U method has been shown effective at screening castings or other small parts. A variety of applications, developed following a typical empirical process, have been demonstrated to show that the method is useful. First, an acoustic response signature is established for a satisfactory part; then deviations of the signature due to different types of defects are catalogued. Inspection consists of interrogating each part or casting on an assembly line to detect those having a deviate acoustic response signature.

The draft standards listed as follows were being written in 1991. These new documents will help to generalize the future development of the instruments and procedures for applications of A/U. It is important at this stage to avoid procedures that apply to a single instrument.

ASTM 403/89-3 (draft): Standard Guide for the Application of Acousto/Ultrasound

The purpose of this tutorial guide is to establish the rationale and basic methodology for the practice of the A/U technique. In addition to the tutorial guide, four practices covering different applications have been started under the tentative titles listed below.

1. Impact damage detection in composites.
2. Adhesive bond evaluation.
3. Wood products integrity.
4. Composite cure monitoring.

ASTM 402/85-1 (draft)—Secondary (Transfer) Calibration of AE Sensors

A draft document is being developed to cover the transfer of sensor calibration performed at NIST according to the primary calibration standard (E 1106) to the user's sensors. The document will address the technical problems of adequate excitation of a sensor for measurement of the pertinent parameters of sensor response to make a comparative measurement traceable to the primary standard.

ASTM Draft: Standard Practice for Assessment of Small Parts or Structures at Restricted Areas of Interest

This recommended practice describes requirements for conducting AE examinations of small parts. It is confined to test objects for which, because of their dimensions and structure, integral monitoring without location facilities or with simple linear location facilities can be considered sufficient. The purpose is to determine the structural integrity of a test object under specific loading conditions. The application of this practice is confined to objects with a typical emission behavior which can be examined in a preceding series of tests made on equivalent parts or a small, permanent installation. Possible applications of this proposed practice are control of production processes, prooftesting after fabrication, and retesting after intervals of service.

ASTM Draft: Test Method for Zone Location Based Acoustic Emission Examination of Metal Storage Tanks

This procedure defines instrumentation requirements, test procedures, and evaluation criteria for zone-location AE testing of above-ground storage tanks to evaluate structural integrity. Design, fabrication, and construction of storage tanks AE tested per this procedure comply with American Petroleum Institute specifications or other applicable codes and standards. This procedure applies to new and in-service tanks constructed of carbon steel, stainless steel, aluminum, and other metals.

Future Needs

International Standards

Many standard-generating organizations are participating in the development of international standards. The International Standardization Organization (ISO) is addressing the very difficult process of achieving a consensus for nondestructive evaluation (NDE) standards developed in the USA, Europe, Japan, and elsewhere. In the past, the world AE community has developed national standards for AE services that usually differ somewhat from country to country in details. A compromise will have to be reached on the details to produce internationally acceptable standards.

Offshore

AE has been evaluated on some offshore applications with limited success. Some interesting problems, such as monitoring the integrity of mooring cables and tension legs on offshore platforms, meet with difficulty in handling the high background noise from wind, waves, sea life, and mechanical noises on the platform. Instrumentation must be hardened against salt corrosion and operation at hundreds of feet in depth.

The work to apply AE on offshore structures is still experimental, but given the lead time required to produce a useful standard, ASTM must be alert to the development of new instruments or procedures that make offshore AE services practicable.

Underwater

Many of the same problems encountered on offshore platforms apply to underwater AE monitoring with the exception that working deeper than 200 feet may be less noisy than surface work. The barnacles and small fishes that create a noisy surface environment live above 200 feet, and the wind and wave activity are not noticeable in deep water. The AE applications to underwater NDE include evaluating welds, monitoring fatigue, and monitoring migration of pipelines or mooring anchors, to name a few. At present, applications of AE underwater are not routine, but standards will be needed in the future.

Aircraft

The use of AE on aircraft goes back to the late 1960s when a small plane was instrumented to evaluate the noise conditions attendant to monitoring structural members such as wing spars. While noise was a problem, high-stress maneuvers did generate increases in AE rates that were deemed significant. Later AE was applied to Australian military aircraft on an experimental basis. More recently, part of the Air Force fleet of C5A transports was instrumented for AE evaluation of wing spars during programmed test-flight sequences.

The technique of using AE for structural integrity in aircraft may be sufficiently advanced to proceed with standards development, but the incentive provided by a participating vendor of AE services is missing.

Conclusions

Standards have been instrumental in generalizing the development of AE services technology. Application and analysis procedures have become independent of specific instruments or AE service vendors. The industrial users of AE services are protected by standards, and the vendors of AE services have credibility through standards.

The public safety also is assured by standards. The consensus process of standard development exposes the techniques and procedures of a developing standard to a level of scrutiny that cannot be achieved within one organization. The resulting standard then is continually refined by review and revision to assure the public safety.

References

[1] Cross, N. O., et al., "Acoustic Emission Testing of Pressure Vessels for Petroleum Refineries and Chemical Plants," *Acoustic Emission, ASTM STP 505*, American Society for Testing and Materials, Philadelphia, 1972, pp. 270–296.
[2] Spanner, J. C., "Acoustic Emission Techniques and Applications," Intex Publishing Company, Evanston, IL, 1974, distributed by the American Society for Nondestructive Testing, Columbus, OH.
[3] Drouillard, T. F., Liptai, R. G., and Tatro, C. A., "Industrial Use of Acoustic Emission for Nondestructive Testing," *Monitoring Structural Integrity by Acoustic Emission, ASTM STP 571*, American Society for Testing and Materials, Philadelphia, 1975, pp. 122–149.

Patrick C. McEleney[1]

Electromagnetic (Eddy Current) Testing

REFERENCE: McEleney, P. C., "**Electromagnetic (Eddy Current) Testing,**" *Nondestructive Testing Standards—Present and Future, ASTM STP 1151*, H. Berger and L. Mordfin, Eds., American Society for Testing and Materials, Philadelphia, 1992, pp. 63–70.

ABSTRACT: This paper will cover the electromagnetic (eddy current) testing method and ASTM Subcommittee E7.07 on Electromagnetic Method and its activities. The basic principles and applications of the test method are discussed. The advantages and disadvantages are noted. Areas of standardization highlighted are terminology, equipment, test procedure, and reference standards. These referenced standards cover sorting, coating thickness, conductivity, and tubular products.

The background of ASTM Subcommittee E7.07 and its standard chronology are highlighted. The subcommittee's growth from three sections in 1958 to its present ten sections is covered in detail. Extensions of the scope to cover the magnetic method for coating thickness, fringe flux, and the electric potential method are noted. Thirteen documents have been developed by the subcommittee, and six new documents are currently in development. Of high interest currently is the work on wire rope examination standards. Many challenges are presented to ASTM E.07 including: (1) developing documents that can replace DoD specifications and standards, and (2) developing sorely needed documents for wire rope inspection. This paper notes a high level of standard document development currently.

KEY WORDS: electromagnetic testing, eddy current testing, conductivity measurement, coating thickness measurement, sorting, tubular products inspection, electromagnetic (eddy current) standards, wire rope inspection, fringe flux, electric potential

A good definition of *electromagnetic (eddy current) testing* has never been developed. So, a summary of the basic principles and applications of the test method will serve to put everything in perspective.

Basic Principles of Test Method

In electromagnetic testing, energy is lost in the specimen by two separate processes: (a) magnetic hysteresis and (b) eddy current flow. In magnetic materials, both effects are present. In nonmagnetic and magnetically saturated materials, the hysteresis effect is absent or suppressed, and the prevalent losses are due to eddy currents. *Saturation* is a term used generally to describe the condition of ferromagnetic material at its maximum values of magnetization.

Eddy currents are induced in a specimen by a time-varying magnetic field usually generated by an alternating current flowing in a coil. The coil configuration may assume a wide variety of shapes, sizes, and arrangements. The coil may surround the specimen (encircling) or may be placed on or near the surface (probe).

Eddy currents are influenced by many characteristics of the material: conductivity, magnetic permeability, geometry, and homogeneity. This fact makes it possible to evaluate many different characteristics of the specimen with a suitable test setup.

[1] Consultant, U.S. Army Materials Technology Laboratory, SLCMT-MEE, Watertown, MA 02172-0001.

Eddy current effects are most pronounced near the surface, with sensitivity for detecting irregularities of composition or structure falling off as depth below the surface increases. Depth of eddy current penetration of an object decreases as test frequency increases. Ferromagnetic metals, such as steel, are generally tested with low frequencies in the range of 1 to 10 000 Hz (10 kHz). Nonmagnetic metals with high conductivity, such as aluminum, are generally tested with frequencies around 100 kHz, while those with lower conductivity, such as titanium, are generally tested with frequencies in the range of 1 to 10 MHz. There are numerous exceptions to these generalities.

Applications of Test Method

General

The electromagnetic (eddy current) method is used for determining surface imperfections. Under appropriate conditions and with proper instrumentation, the method has been used to:

1. Detect discontinuities such as, but not limited to, seams, laps, slivers, scabs, pits, cracks, voids, inclusions, and cold shuts.
2. Sort for chemical composition on a qualitative basis.
3. Sort for physical properties such as hardness, case depth, and heat damage.
4. Measure conductivity and related properties.
5. Measure dimensions such as the thickness of metallic coatings, plating, cladding, wall thickness or outside diameter of tubing, corrosion depth, and wear.
6. Measure the thickness of nonmetals when a metallic backing sheet can be employed.

Equipment

The electronic apparatus shall be capable of energizing the encircling coil or probe with alternating currents of suitable frequency and amplitude and shall be capable of sensing the electromagnetic response of the sensors. Equipment may include a detector, phase discriminator, filter circuits, modulation circuits, magnetic saturation devices, recorders, and signaling devices as required by the application.

The encircling or probe coil assembly used shall be capable of inducing current in the part and sensing changes in the electric and/or magnetic characteristics of the part.

A mechanical device capable of passing a part (such as a tube) through the encircling coil or past the probe may be used. It shall operate at uniform speed with minimum vibration of the coil, probe, or part and maintain the article to be inspected in proper register or concentricity with the probe or encircling coil. A mechanism capable of uniformly rotating or moving the part or the probe may be required.

An end effect suppression device, a means capable of suppressing the signals produced at the ends of tubes, may be used.

Reference standards are required to adjust the sensitivity of the apparatus.

Typical Examples of Equipment Variations for Different Applications

1. Equipment using impedance plane analysis and operable over a range of test frequencies from 1 Hz to 10 kHz has been used to sort carbon steel fixtures involving different compositions and/or different heat treatment conditions. A unique advantage of this instrument is that it is possible to quickly determine the optimum frequency of performing a given test.

Similar equipment has been calibrated to indicate conductivity, hardness, case depth, and dimension.

2. Equipment using a single coil to scan the surface has been used to detect and indicate the depths of seams, cracks, laps, slivers, and similar surface imperfections in bars, rounds, billets, and tubular products. The sensitivity of this equipment depends on the surface condition of the product under test. On a hot-rolled surface with thin, tightly adherent scale, seams as shallow as 0.010 in. (0.025 cm) are reliably evaluated. Products with heavy or broken scale should be cleaned by grit blasting prior to testing. Under more favorable (smoother, less scale) surface conditions, seams as shallow as 0.005 in. (0.013 cm) have been evaluated. On polished (ground) surfaces, seams and cracks as shallow as 0.001 in. (0.003 cm) have been detected.

3. Equipment using differential test coils has been used to detect imperfections in carbon steel tubular and bar products. Test frequencies ranging from 400 Hz to over 20 kHz have been used. At the lowest test frequencies, and with the use of magnetic saturation, defects have been reliably detected (outer surface, inner surface, or subsurface) in the walls of tubular products with wall thicknesses as great at 5/8 in. (1.59 cm). When testing at frequencies as low as 400 Hz, the test speed is limited to about 100 ft/min (3.05 km/min). When higher test frequencies are used, the test speed can be correspondingly increased. Higher test frequencies can be used for testing products with thinner walls and higher resistivity.

Advantages/Disadvantages of Test Method

Advantages

One of the advantages of electromagnetic (eddy current) equipment is that it lends itself to automatic operation for regularly shaped parts. Tubing is ideal for automatic inspection, and speeds to 500 ft/min (15.24 km) were permitted in the 1967 version of ASTM Standard E 215. Speeds far in excess of 500 ft/min (15.24 km) were employed a few years later.

Other advantages include the fact that intimate contact or couplant is not required between the coil and the material. The test is versatile. Special coils can easily be made. Electric circuit design permits selective sensitivity and function. The test is sensitive to surface or near surface inhomogeneities. No special operator skills are required. The test is low in cost, especially in high-speed, automatic operation. Another advantage to this method is that it has permanent record capability for symmetrical parts.

Manual systems which are small, simple, and inexpensive are common in applications involving large or irregularly shaped objects.

Disadvantages

Masked or false indications can be caused by sensitivity to variations such as part geometry, lift-off, and permeability. The method is only useful on an object containing a conductive material to establish the electric and magnetic field. The shallow depth of penetration is a disadvantage in certain applications. Reference standards are required and are difficult to make. Edges, speed, temperature, magnetic history of the part, and surface condition can affect the test. It is also sensitive to many other variables besides the one of interest, and these must be controlled. Sensitivity varies with depth. Response is frequently comparative and not quantitative.

Areas of Standardization

Terminology

Although many of the other methods subcommittees in ASTM Committee E-7 on Nondestructive Testing had been in existence for many years before E7.07 was formed—the Electromagnetic (Eddy Current) Subcommittee was among the first to publish a complete glossary, ASTM E 268. This represented a significant achievement as there was very little in the way of existing electromagnetic glossaries at that time. The literature and foremost authors had developed a wide range of terms and definitions, so much so that one might have had great difficulty in following a recommended practice without a glossary, even though experienced in the method. Today one may experience some difficulty with certain authors, but the language of electromagnetic standards is clear and concise.

Equipment

Electromagnetic equipment can be large, elaborate, and expensive when multiple stations and materials handling sections are included such as used on sheets and plates. Manual systems which are small, simple, and inexpensive are also common in other instances such as those used with large irregularly shaped objects.

Vector-sensitive instruments operate on the impedance plane principle. The frequency range of these instruments is from 100 Hz to 6 MHz. This type of operation considers both the amplitude and phase of the eddy currents. This allows one to optimize the instrument response for a selected material variable, while minimizing response to another variable, such as probe spacing.

Methods

The versatility and wide range of applications for the electromagnetic (eddy current) method would indicate a rather broad area for standardization. This is the case as is shown in the many documents under the jurisdiction of E7.07.

The effect of the characteristics of the specimen on the eddy currents may be studied in a number of different ways. A characteristic to be studied is related to a change in the amplitude, distribution, or phase of the eddy currents, or some combination of these three. These changes in the eddy currents are reflected as changes in the exciting coil or auxiliary coils so located as to be sensitive to the desired eddy currents. These changes may be measured as voltage differences, current differences, phase differences, or changes in the impedance of the coil(s).

The coils and the instrumentation can be arranged to measure a given characteristic directly, or they may be used as a comparator. In the latter case, the measurement is the difference between the characteristics of the specimen and a similar part of known or acceptable characteristics. Such a measurement can also be made to determine the difference between various segments of the same specimen. With the best instrumentation, it is sometimes difficult to separate effects of the characteristics to be measured from effects of other characteristics. The success of the eddy current test depends on: (1) proper coil design and arrangement; (2) selection of the proper test frequency and analysis circuit; (3) use of proper magnetic field strength; (4) optimization and maintenance of electromagnetic coupling between the coil and specimen, and (5) selection of the most suitable reference standards.

Reference Standards

Sorting Standards—In sorting, using the absolute coil (encircling) method, a known acceptable calibration standard and a known unacceptable standard are required. When using the comparative coil (encircling) method, usually two known acceptable specimens of the specimen tested and one known unacceptable specimen are required. For a three-way sort, it is best to have three calibration standards, two of which represent the high and low limits of acceptability for one group or one each of the two unacceptable groups. The third standard represents the acceptable lot of material.

Coating Thickness Measurements Standards—Calibration standards of uniform thickness are available in either of two types: foil or coated substrate.

Conductivity Standards—These are of two types:

1. *Primary standards.* Primary standards are those standards which have a value assigned through direct comparison with a standard calibrated by the National Institute of Standards and Technology or have been calibrated by an agency which has access to such standards. The primary standards are usually kept in a laboratory environment and are used only to calibrate secondary standards.
2. *Secondary standards.* Secondary standards are those standards supplied with the instrumentation or standards constructed by the user for a specific test.

These standards are used to calibrate the instrumentation during most testing of materials.

Standards for Tubular Products—The standard used to adjust the sensitivity of the apparatus shall be free of interfering discontinuities and shall be of the same nominal alloy, heat treatment, and dimensions as the tubular products to be examined. It shall be of sufficient length to permit the spacing of artificial discontinuities to provide good signal resolution and be mechanically stable while in the examining position in the apparatus. Artificial discontinuities placed in the tube shall be one or more of the following types:

1. *Notches.* Notches may be produced by electric discharge machining (EDM), milling, or other means. Longitudinal, transverse notches, or both may be used. Orientation, dimensions, configuration, and position of the notches affect the response of the eddy current system.
2. *Holes.* Drilled holes may be used. They are usually drilled completely through the wall. Care should be taken during drilling to avoid distortion of the specimen and hole.

Background of Subcommittee and Standards Development Chronology

The baby in the group of surface methods (liquid penetrant, magnetic particle, and electromagnetic methods) is ASTM Subcommittee E7.07 on the Electromagnetic (Eddy Current) Method. The Electromagnetic Committee had its first meeting in Boston, Massachusetts on 26 June 1958. The subcommittee still boasts several of the original group who are still active: P.C. McEleney, Chairman; Robert W. McClung; and Robert G. Strother. Another member still active in ASTM Committee E7 is Arnold Greene.

Temporarily chairing the group in Boston at the organizational meeting was Hamilton Migel, then chairman of E7.03. One of the major elements in need of electromagnetic standards at the time was the nuclear industry, and a large element of the membership was drawn from this group. Among them was Robert Oliver, secretary pro-temp of the organization meeting from Oak Ridge National Laboratory. E7 Chairman (1958) Jim Bly appointed John W. Allen of Oak Ridge National Laboratory the first permanent chairman of E7.07. W. A. Black, director of research for Republic Steel, was first permanent secretary.

Howard Bowman, director of research for Trent Tube, was appointed chairman of Section .01 on Nonmagnetic Tubing.

There were two areas relating to eddy current testing in which other ASTM committees were interested and which were proposed at this organizational meeting.

1. *Conductivity measurement*—ASTM Committee B-7, Subcommittee .03 had been considering specifying the eddy current method for conductivity measurements of aluminum.

2. *Testing nonmagnetic bars and tubes*—ASTM Committee B-5 had given some consideration to specifications relating to eddy current inspection of copper and copper alloy tubes. There was also considerable interest in specifications relating to eddy current inspection of stainless steel tubing for atomic energy applications.

A questionnaire was sent out by the chairman, Subcommittee E7.07, in September 1958, in which three task groups were noted.

1. Task Group I for the inspection of tubular products fabricated from metals and alloys other than those that are strongly magnetic.
2. Task Group II to compile a glossary of terms related to the work of the committee.
3. Task Group III to work with American Standards Association Special Committee C-31 on a code for pressure piping.

At the 5 February 1959 meeting in Pittsburgh, Task Group 4 was established to write a standard practice covering the measurement of electrical conductivity by the electromagnetic (eddy current) method. In addition, a Task Group 5 was appointed to consider preparation of a recommended procedure for the use of eddy currents in the measurement of thickness. It is interesting to note that R. B. Oliver (deceased), J. Callan (deceased), and P. C. McEleney showed up at this meeting with proposed glossaries of terms.

Little support was shown for Task Group 3, although activity continued. Before the June meeting, Task Group 4 chairman, H. Migel, noted that Committee B-1 had submitted a similar specification for letter ballot which had been prepared by a B-1 task group headed by G. W. Stickley of Alcoa Research Laboratories. This he sent out to E7.07 for review and comments. These two groups eventually were inactivated and disbanded. In June 1959, Task Group .03 was dropped and in January 1960 Task Group .04 terminated. However, a Task Group .06 covering magnetic tubing was formed at the January 1961 meeting.

In January 1960 the scope of the subcommittee was expanded to cover magnetic methods of measuring coating thickness.

In June 1962 H. Bowman resigned as chairman of Section .01 and was replaced by R. B. Socky of General Electric Co. In January 1963 John Allen resigned. He was succeeded in March 1963 by W. A. Black. P. C. McEleney was appointed secretary.

In December 1964 a joint task group was formed of A5 and E7 members to develop a single document combining E 216 and A 464, the committees' respective coating thickness measurement standards. The task group was chaired by Fielding Ogburn of the National Bureau of Standards. This work resulted in a new document, E 376–68, replacing E 216 and A 464. It is interesting to note that A 219, Tests for Local Thickness of Electrodeposited Coatings (see 1970 Book of ASTM Standards, Parts 3 and 7), was discontinued about this time.

Other events:

ASTM A 464: Recommended Practice for Use of Magnetic Type Instruments for Measurement of Thickness of Hot-Dip Zinc Coatings on Iron and Steel (withdrawn).
ASTM E 216: Recommended Practice for Measuring Coating Thickness by Magnetic or Electromagnetic Methods (withdrawn).
In May 1965 W. A. Black resigned, succeeded by N. H. Cale of Anaconda American Brass Co. in August 1965.

This is how it all began. In succeeding years the scope was expanded to cover the magnetic flux leakage test method for detection of outer surface, inner surface, and subsurface discontinuities in ferromagnetic steel products.

In December 1969 N. H. Cale was succeeded as chairman by Patrick C. McEleney of the Army Materials and Mechanics Research Center. Walter H. Rolfe of the Magnaflux Corporation was appointed secretary. R. B. Socky was succeeded by R. I. Buckley of Texas Instruments as Chairman of Section .01. He in turn was succeeded by Matthew Dashukewich of the Magnaflux Corporation.

One of the highlights of the earlier years was an extensive round robin on three sets of non-magnetic tubing (copper, aluminum, and stainless steel). There were many participants in this round robin, and the results were finalized and distributed by Art Moughalian of Phelps Dodge in February 1968 in a herculian effort. The study provided much useful information on drilled holes, filed notches, and elox notches.

New sections were formed: (1) E7.07.08 for sorting ferrous metals and nonferrous metals; (2) E.07.04 for measurement of electrical conductivity by eddy current methods; (3) E7.07.09 for detection of outer surface, inner surface, and subsurface discontinuities using magnetic flux leakage fields employing either the residual magnetic fields or active magnetic fields; (4) E7.07.07 for high-temperature eddy current applications; (5) E7.07.03 on electric potential, and (6) 07.07.10 on wire rope applications.

A review of the documents under the jurisdiction of E7.07 unfolds the following background information:

E 215: Standardizing Equipment for Electromagnetic Testing of Seamless Aluminum-Alloy Tube—originally issued in 1963.

E 243: Electromagnetic (Eddy Current) Testing of Seamless Copper and Copper-Alloy Tubes—originally published in 1967.

E 268: Standard Definitions of Terms Relating to Electromagnetic Testing—originally published in 1965.

E 309: Eddy Current Examination of Steel Tubular Products Using Magnetic Saturation—originally published in 1966.

E 376: Measuring Coating Thickness by Magnetic—Field or Eddy Current (Electromagnetic) Test Methods—originally issued in 1968. This replaced E 216 and A 464.

E 426: Electromagnetic (Eddy Current) Testing of Seamless and Welded Tubular Products, Austenitic Stainless Steel, and Similar Alloys—originally published in 1971.

E 566: Electromagnetic (Eddy Current) Sorting of Ferrous Metals—originally published in 1976.

E 570: Flux Leakage Examination of Ferromagnetic Steel Tubular Products—originally published in 1976.

E 571: Electromagnetic (Eddy Current) Examination of Nickel and Nickel Alloy Tubular Products—originally published in 1976.

E 690: In Situ Electromagnetic (Eddy Current) Examination of Nonmagnetic Heat Exchanger Tubes—originally published in 1979.

E 703: Electromagnetic (Eddy Current) Sorting of Nonferrous Alloys—originally published in 1980.

E 1004-84: Test Method for Electromagnetic (Eddy Current) Measurements of Electrical Conductivity.

E 1033-85: Practice for Electromagnetic (Eddy Current) Examination of Type-F Continuously Welded (CW) Ferromagnetic Pipe and Tubing Above the Curie Temperature.

E 1312-89: Practice for Electromagnetic (Eddy Current) Examination of Ferromagnetic Cylindrical Bar Product Above the Curie Temperature.

New Standard Practice for Nondestructive Examination with Electric Potential Techniques.

New Standard Practice for Electromagnetic Examination of Wire Rope.

Subcommittee ASTM E7.07 is currently set up with the following sections:

E7.07.01: Nonferrous Application
E7.07.02: Terminology
E7.07.03: Electric Potential
E7.07.04: Conductivity Measurement
E7.07.05: Thickness Measurement
E7.07.06: Ferrous Applications
E7.07.08: Sorting
E7.07.09: Fringe Flux
E7.07.10: Wire Rope Application

Many new document(s) are being developed. The many documents already developed are being maintained. New developments and needs are being reviewed and action taken where needed. The workload is large, the workers few.

The future: The work is not completed, far from it—we have barely just begun. Some areas were mentioned by speakers.

1. Eddy currents for temperature measurement (back where we began in the 1880s)—Len Mordfin.
2. Probe coil evaluation—Bernie Strauss.
3. Eddy current examination of welds—E. Borloo.

Other areas which might be developed:

1. Equipment certification.
2. Pulsed eddy currents.
3. Remote field eddy currents.

The late Dr. McMaster at the international meeting on nondestructive testing in Las Vegas a few years back suggested many unexploited areas including coil design. NIST recently came out with standards for eddy current testing.

These are just another advance in this area—much more is needed. Standards for wire rope inspection are sorely needed.

The book is not closed on electromagnetic (eddy current) examination. We have just touched the tip of the iceberg.

Robert W. McClung[1]

The Challenge of Standards for Emerging Technologies

REFERENCE: McClung, R. W., "**The Challenge of Standards for Emerging Technologies,**" *Nondestructive Testing Standards—Present and Future, ASTM STP 1151,* H. Berger and L. Mordfin, Eds., American Society for Testing and Materials, Philadelphia, 1992, pp. 71–74.

ABSTRACT: The preparation of standards for emerging NDT methods may sound paradoxical with the rapid changes normally associated with new and emerging technology. However, consensus documentation of terminology and basic methods of performing tests of NDT equipment can do much to aid in the emergence and acceptance of such technology. Subcommittee E7.10 on Other NDT Methods was established to serve as an "umbrella" group and home for standards activities for emerging or other technology that has not yet grown sufficiently large to justify an independent subcommittee. Current activities include sections on thermoelectric sorting of metals, optical holography, and infrared methods. In addition, there are task groups on NDT reliability and methods for metal verification, identification, and sorting.

KEY WORDS: nondestructive testing, nondestructive evaluation, infrared, thermoelectric, optical holography, standards

As discussed in several of the companion papers in this publication, the activities of ASTM Committee E7 on Nondestructive Testing include preparation of standards on various aspects of nondestructive testing (NDT). The major part of the membership of the committee and attendant action is on long-time, well-established, and recognized technologies such as radiography, liquid penetrants, ultrasonics, and others. With such a broad historical base, for many of the activities there are established industrial techniques and many potential participants on which to draw for ASTM standardization. For such, the major (and not trivial) business is to establish consensus documentation. As new, more complex, or sophisticated techniques and equipment are developed in the "old-line" methods of nondestructive testing, the standardization also becomes more complex with fewer available participants and industrial techniques. But what about the "smaller" techniques which may represent new emerging methods and which do not enjoy the luxury of many participants or established industry practices and techniques?

In 1981, the leadership of Committee E7 recognized that the existing technical subcommittee structure did not provide a logical "home" for NDT technologies that were not an integral part of the recognized standardized methods of NDT. Since the mid-1970s there had been an administrative subcommittee on Special NDT Methods with a charter to investigate new methods and techniques, to serve as an educational arm of the committee through organization and sponsorship of seminars, and to make recommendations for new standards activities. However, in some instances, there was no readily identifiable technical subcommittee for

[1] Consultant, Oak Ridge National Laboratory, P.O. Box 2008, Oak Ridge, TN 37831-6158. Oak Ridge National Laboratory is operated by Martin Marietta Energy Systems, Inc. under contract DE-AC05-840R21400 with the U.S. Department of Energy.

jurisdiction! To address this problem, technical Subcommittee E7.10 on Other NDT Methods was formed. (Several years later, at the recommendation of the subcommittee, the title was changed to Emerging NDT Methods to be a more accurate representation of the charge and activities.)

Organization of Subcommittee E7.10

In contrast to the existing technical subcommittees of E7, with the entire thrust and sub-structure organized to address a single NDT method, it was recognized that E7.10 would provide a technical umbrella under which many diverse methods could be represented and function for standardization activities. The methods could involve a small part of the NDT world with relatively few practitioners and the need for only a few standards (and no expectation for major proliferation or growth). On the other hand, the method could be a new advance or a new entry into NDT standardization (that did not fit other existing subcommittees) with potential for major growth from a small beginning. In either case, E7.10 was intended to provide a home, either permanently or on an interim basis, until the number of participants and volume of activity justified a larger organizational role (e.g., subcommittee or even committee status).

The initial organization of Subcommittee E7.10 established four sections to meet recognized needs at that time. These were:

1. E7.10.01 on Thermoelectric Sorting of Materials.
2. E7.10.02 on Optical Holography.
3. E7.10.03 on Visual/Optical Methods.
4. E7.10.04 on Infrared Methods.

Later, based on recognized technical needs, two task groups were established for initial standardization activities pending recognition of the ultimate organizational structure based on participation, growth, and the need for standards. The first was a task group on Material Verification, Identification, and Sorting. Some time later, a second task group on NDT Reliability was formed.

Generic Problems and Challenges

Despite the diversity and uniqueness of each of the Committee E7.10 sections and task groups, there were several common problems and challenges in the formation, staffing, and work organization and implementation. Although the basic technologies were not new, there was a lack of prior documentation or standardization. Each step toward standardization was original with little or no written precedents. The number of individuals involved in the development and/or application of the technology was generally small. The normal changes in technical interest due to job or company changes further exacerbated the difficulties of section or task group staffing. This resulted in every advance in standardization being the product of a small, dedicated core of individuals. (The latter is, in fact, no different from the experience of many committees and subcommittees, but the pool of potentially available workers for E7.10 is much more limited.) A major difficulty toward communication and documentation in technology with no standards was the lack of common understanding of the language. Therefore, one of the first steps for some of the groups was definitions of terms. These not only assisted in further documentation, but provided a common, solid foundation (and at the same time improved overall credibility) for all those using the technology and for the rest of the technical community.

As new standards were addressed (where none had been before), it was necessary to be concerned about the readiness of the technology for standards. As with all technology, the goal was to identify consensus good practice for performance of a technique or procedure without stifling or discouraging beneficial developments or advances.

Despite these difficulties, the active semiautonomous section and task groups have each made significant progress toward meeting the identified unique and specific needs for preparation of standards peculiar to the technology. The activities and products through the time of writing of this paper will be briefly described.

Activities of Sections of E7.10

E7.10.01 on Thermoelectric Sorting of Materials, one of the original sections, was the first to address a standard for its subject technology. Commercial instruments had been available in industry for a number of years to allow separation of metals (e.g., according to constituents or heat treatment) based on the thermoelectric effect (the Seebeck coefficient). However, there had been no consensus documentation on the proper procedures to apply the equipment and interpret the results. Interested individuals in industry who were also members of Committee E7 had already begun a draft procedure for the technology. However, there was no logical home for the action within the then-existent E7 structure. Therefore, when the new organization structure (and E7.10.01) was established, the new section had a head start. The draft standard proceeded through the consensus system and was approved as ASTM Practice for Thermoelectric Sorting of Electrically Conductive Materials (E 977). It has stood the test of time and was reapproved without revision in 1989. Current interest and activity within the Section is on a round robin basis to evaluate the method for determination of hardness in selected metals.

E7.10.02 on Optical Holography (for NDT) has developed slowly because of changes in equipment and applications for this rapidly evolving technology. In addition, as noted earlier, the number of involved individuals with an interest in standards is very limited. One of the earliest organized needs was for consensus definitions of terms. After early efforts to identify and compile terms to be defined, a lengthy, laborious, and diligent effort was successful in establishing a consensus standard which was approved through the required balloting process. The document was integrated as a subdivision into the overall E7 glossary document ASTM Terminology for Nondestructive Examinations (E 1316). Additional terms are being identified and defined and procedural documents are being addressed.

E7.10.03 on Visual/Optical Methods, although identified in the organizational structure, has never been activated for lack of dedicated leaders and workers with adequate interest in promotion and preparation of standards. E7.10 is continuing to investigate the needs and to sponsor educational activities (e.g., in fiber optics) that may lead to Section activation.

E7.10.04 on Infrared Methods has been the most productive of the sections. An extensive list of terms related to NDT by infrared has been identified and defined with a consensus document, E 1149, Standard Terminology Relating to NDT by Infrared Thermography, being issued in 1987. This has now been incorporated into the overall E7 glossary document ASTM E 1316. Additional terms are being addressed as needed and added to the list of definitions. The activities toward procedural documents has been directed to standards for checking the performance of infrared systems. The intent is to provide common ground for sellers, purchasers, and users of infrared equipment to discuss and evaluate the capabilities of individual infrared systems. The first such document to proceed from draft stages to consensus-approval standard was ASTM Test Method for Minimum Resolvable Temperature Difference for Thermal Imaging Systems (E 1213) issued in 1987. The next issued standard was E 1311, Test Method for Minimum Detectable Temperature Difference for Thermal Imaging Systems, in

1989. Current activities are addressing an additional standard on noise equivalent temperature difference.

Activities of Task Groups of E7.10

As noted earlier, new activities in E7.10 may be assigned to a task group rather than being made a part of the permanent organization as a section. The temporary status is accorded pending recognition of the proper organization based on growth and activity.

A chronic problem with variations in intensity, severity, and recognition has been the identification, verification, or sorting of materials, particularly metals. E7.10.01, discussed earlier, is one method of addressing the difficulty. An industrial survey (with a questionnaire) was conducted by E7.10 to elicit opinions about available relevant methods, the needs or desirability for standards, and (if standards are needed) identification of candidate workers. The results showed several commonly recognized methods, the need for standards, recommendations to start with methods for metals, and several who expressed an interest in participation. A task group was then formed on metal verification, identification, and sorting.

The difficult task of developing standards was made more onerous by the limited number of working volunteers and the fact that some of the technical experts in specific methods of evaluation were limited in their knowledge of other NDT technologies and standards writing procedures. The initial selected goal was the preparation of a standard guide that would identify many of the methods for material identification and sort and briefly discuss the applicability and capability. The document, "Standard Guide for Metals Identification, Grade Verification, and Sorting," is currently proceeding through the ballot process. After successful compilation of the new standard guide, decisions and recommendations will be made about the next needed standards as well as current consideration about the organizational structure for the activity.

An increased emphasis on the quantitative capability of NDT, enhanced by growing use of fracture mechanics analysis and probabilistic design and risk assessment, has raised questions about the consistency or reliability of NDT procedures. Studies by the aerospace and nuclear industries demonstrated significant variations in the performance and results of nondestructive examinations. These have led to a recognition of the need for improved methods to evaluate the capability and reliability of NDT. Early attempts to address the problem were, in general, limited in scope (e.g., to determine probability of detection of flaws) to the near-term needs of the organization. A Task Group on NDT Reliability has now been established in E7.10 to provide a consensus view of methods of measuring NDT reliability that will be generally useful to all industry. The initial standardization activity is based on a draft Air Force report compiled by several industry participants with emphasis on aircraft engines.

Summary

Subcommittee E7.10 on Emerging NDT Methods is a unique group serving as an umbrella organization to provide administrative support for preparation of standards in small, diverse methods of NDT. The difficult task of generating original standards with limited precedents is made more difficult by the limited population of technical experts from which to draw (and these may have limited knowledge of the standards process). Despite the problems, excellent progress and productivity is being made by a small group of dedicated individuals.

William C. Plumstead¹ and Claude E. Jaycox²

Standards for NDT Laboratories

REFERENCE: Plumstead, W. C. and Jaycox, C. E., "**Standards for NDT Laboratories,**" *Nondestructive Testing Standards—Present and Future, ASTM STP 1151,* H. Berger and L. Mordfin, Eds., American Society for Testing and Materials, Philadelphia, 1992, pp. 75–81.

ABSTRACT: In 1976, after about ten years of debate and repeated ballots, the ASTM Practice for Determining the Qualification of Nondestructive Testing Agencies (E 543) gained final approval and was published. The document provides the minimum requirements considered essential to the proper organization, administration, and operation of commercial and in-house agencies providing nondestructive examination services. It represents an industry consensus for the minimum practices expected of a qualified nondestructive testing agency. It includes guidelines for equipment maintenance and calibration for the most frequently used NDT methods, personnel qualification and certification, and minimum requirements for an effective quality manual.

Subsequent documents to supplement E 543 have been developed that offer detailed guidelines for a laboratory quality control system (ASTM E 1212) and for a survey (audit) checklist (ASTM E 1359) for use by evaluators of nondestructive testing agencies.

Nondestructive testing agencies that meet the requirements of ASTM E 543 and ASTM E 1212 can be expected to offer consistent quality performance and reliable examination results. These ASTM documents are being specified more frequently than in past years, but, unfortunately, are not being applied as they should for the best results. In many cases, the lowest bidder is assumed to be qualified, but the qualifications are not verified. Too often, when evaluations of nondestructive testing agencies are conducted, individuals are used who are not technically knowledgeable in the field. Frequently, these evaluations are conducted after the award of a contract, when changes are very difficult. Increased use of these ASTM standards by industry combined with competent evaluations of the quality of an agency's work and its technical and management skills will result in increased levels of confidence on the part of the users of products inspected and examined by these agencies. These higher levels of confidence will improve the acceptance of U.S. products worldwide.

KEY WORDS: agency, labs, surveys, checklists, quality manual

Nondestructive testing agencies (laboratories) both commercial and in-house offer a variety of services to support construction, fabrication, and manufacturing. The laboratories or agencies that provide these services range widely in their ability to perform specific methods, applications, and techniques. Wide-ranging differences in organization, size, equipment, and personnel exist among agencies providing nondestructive examination services. How can we determine in advance, with reasonable assurance, that a nondestructive testing agency has at least minimum qualifications to provide accurate and reliable results?

Standards

Subcommittee E07.09 on Materials Inspection and Testing Laboratories was the first official subcommittee in ASTM established to work exclusively on documents relating to the quality of the organizations performing ASTM inspections and examinations. It struggled for many

¹ Principal quality engineer, Fluor Daniel, Inc., 100 Fluor Daniel Drive, Greenville, SC 29607-2762.
² President, Municipal Testing Laboratories, Inc., 102 New South Rd., Hicksville, NY 11801.

years to achieve its first industry consensus standard dealing with NDT laboratory qualifications. The ASTM Practice for Determining the Qualification of Nondestructive Testing Agencies (E 543) was first published in 1976. This practice establishes minimum requirements for agencies performing nondestructive examination. It is used to assess the capability and abilities of NDT agencies and could be used as a basis for developing an accreditation procedure.

ASTM E 543 mandated a written quality control manual and a system of process control as an essential part of a quality organization. Because many agencies did not have the capabilities or background to develop these documents without assistance, Subcommittee E07.09 developed a consensus standard to assist them. ASTM E 1212, Practice for Quality Control Systems for Nondestructive Testing Agencies, describes the general requirements to establish and maintain a quality control system and quality control program for agencies engaged in nondestructive examination. An agency is expected to use the document as guidance and to add its specific and unique requirements.

After developing these two documents, the subcommittee responded to complaints of inconsistent and improper evaluations and audits of nondestructive testing agencies by developing the ASTM Standard Guide for Surveying Nondestructive Testing Agencies (E 1359). It is a guide designed to establish areas for review and provides a uniform format for use in determining the technical competence of a nondestructive testing agency. The document as it presently exists is "bare bones." Future subcommittee work will "flesh it out" and add more comprehensive guidelines to assist industry.

All the standards in the world will not improve quality unless used. Presently, the effective utilization of the preceding documents needs vast improvement. While specified more frequently in the past few years, too often they are not being used to the best advantage. In many cases, the lowest bidder is assumed to be qualified, but qualifications are not verified. Too often, when evaluations of nondestructive testing agencies are conducted, individuals are used who are not technically knowledgeable in the field. Frequently, these evaluations are conducted after the award of a contract, when changes are very difficult. Increased use of these ASTM standards by industry, combined with competent evaluations of the quality of an agency's work and its technical and management skills, will result in increased levels of confidence on the part of the users of these agencies. These higher levels of confidence will improve the acceptance of U.S. products worldwide.

A brief overview of each document will identify the salient portions and describe the intended application.

ASTM E 543

ASTM Practice for Determining the Qualification of Nondestructive Testing Agencies (E 543) is applicable where the systematic assessment of the competence of a nondestructive testing agency by a user or other party is desired. It explains the significance and use of the standard and provides a description of the responsibilities, duties, and quality levels expected of nondestructive testing agencies.

Pertinent sections of E 543 are:

Organization of the Agency

The information concerning organization of the agency including company ownership, names of company officers and directors, and organizational affiliates must be documented.

A functional description should describe operational departments and support departments and services.

A brief history of the agency and a description of facilities, capabilities, and services must be provided and includes the types of users of the agency's services.

A list of applicable dates of qualifications, accreditation, and recognition of the agency by others should be included in this documentation of the organization of the agency.

Laboratory Procedure Manual

A laboratory quality manual is required to address the agency's quality control program. This is where the organization outlines its purpose, the hierarchy of responsibilities, control of purchasing, training programs, and quality assurance. Equipment maintenance and calibration must be addressed in addition to a records and documentation section.

Process Control (Operational Procedures)

Procedures provide details for the consistent performance of the various nondestructive testing methods to meet specific codes and client specifications. The procedures should be specific to a particular method and may be specific to a particular application of a method. As an example, an ultrasonic procedure may be specific to the examination of bolting because of specialized technique considerations. Nondestructive examination procedures are required to specify the equipment to be used, calibration requirements, personnel qualification requirements, the details of the method application, acceptance criteria, and reporting requirements. A portion of the procedure may be dedicated to delineation of a step-by-step instruction for a technician to perform the nondestructive examination application.

Personnel Qualification

The organization providing nondestructive examination services must have a personnel qualification and certification program to establish the qualification requirements for employees to perform their specific assignments.

Most agencies offering nondestructive examination services in the United States have a personnel certification program that conforms to the American Society for Nondestructive Testing (ASNT) Recommended Practice No. SNT-TC-1A for the Qualification and Certification of Personnel in Nondestructive Testing. Earlier this year (1991), the ASTM E-7 Committee approved a policy recognizing personnel certification standards other than SNT-TC-1A. The Committee realized that several industry segments and the military reference such documents as MilStd 410 and NAVSEA 250-1500-1 for nondestructive testing qualifications. Additionally, ASNT has published an ANSI approved Standard for the Qualification and Certification of Personnel in Nondestructive Testing (ASNT-CP-189-1991). The new ASNT standard is more stringent than the Recommended Practice No. SNT-TC-1A and provides specific requirements without the flexibility of a recommended practice. ASTM E 543 will be amended, and future revisions of other ASTM E-7 documents will adopt the new wording to allow for other certification programs when appropriate; however, SNT-TC-1A remains the preferred reference. This should reduce potential conflict when using the E 543 standard where personnel certification programs other than SNT-TC-1A are required.

All these programs are very similar, providing the details of education, training, experience, and examinations necessary for qualification to specific levels of capability and responsibility.

Equipment for Nondestructive Testing

This section of E 543 must contain a detailed inventory, listing available equipment. The agency responsible for nondestructive examination of material should be equipped with, or have access to, equipment applicable to the processes used.

Each ASTM E-7 method subcommittee contributed a description of such considerations as equipment, capability, calibration, reference standards, monitoring, and processing requirements as appropriate for the particular method. Every equipment section is intended to be educational, although they do contain some mandatory requirements.

Calibration must be performed to a written procedure that includes calibration standards. Calibration of each machine must be documented with a record or an affixed sticker. Calibration date and due date and the name of the individual who did the calibration should be included in the record.

Equipment Operation and Technique File

Each type of equipment in use shall have a complete manual of all items necessary to operate and maintain the equipment according to applicable codes and specifications. The manual should include maintenance procedures and schedules for each type of equipment.

A technique file should be maintained for each type of examination for the guidance of a technician. This section should provide step-by-step preparation of material for examination, acceptance criteria, control of essential variables, and recording of examination results.

Records and Documentation

The internal process forms or job record forms should be filed with a written report to the client and become a part of the permanent record. The report should include the order number, specification, examination procedure, part identification, customer, and results of the examination. The reports should be signed by the technician performing the work and by a Level II or Level III person. A procedure for Level III auditing of reports should be included.

Specification File

The company should maintain an orderly file containing all the codes, specifications, and amendments under which it is performing work.

E 1212

The ASTM Practice for Quality Control Systems for Nondestructive Testing Agencies (E 1212) describes a standard practice for the establishment and maintenance of a quality control system and a quality control program for agencies engaged in nondestructive examination. The practice outlines and describes the procedures for establishing and maintaining a program for quality control and its continuation through calibration, reference samples, standardization, and examination plans and procedures. The basic quality control system requirements must be documented and encompass the quality policy, planning and administration, organization, resources, and an evaluation system.

Quality Statement, Planning, and Administration

A quality statement shall describe management's specific intention and policy with respect to quality. Major quality objectives and parameters should be approved by the chief executive officer, and periodic organizational audits should be conducted to assure adherence to quality policies.

Programs for planning for each new or modified process or test method should define those characteristics that will be controlled to provide services that comply with defined requirements.

The quality policy and system must be documented and be in an accessible form such as a quality manual or series of manuals. The system's documentation should include its purpose, the organizational outline and hierarchy of responsibilities, control of purchasing, training programs, and quality assurance. Equipment maintenance and calibration must be included along with a records and documentation section.

Human Resources of an Agency

Those aspects of a quality system where the work of the employees will affect the quality of products shall be identified with actions taken to control these aspects.

Duties and responsibilities of personnel shall be identified in job descriptions. Employees shall be selected based on capability and experience or the potential to qualify fully for the job. A training program must be maintained to ensure that employees develop and retain the necessary skill and competence. The training, qualification, and certification of nondestructive testing personnel must meet SNT-TC-1A as a minimum.

Physical Resources of the Agency

The agency shall generally describe its facilities and provide an inventory of its relevant physical resources. A system of written procedures is required for each test method or service performed. An inventory of reference materials including pertinent standards, technical publications, and specifications and amendments should be established and maintained current.

Quality Control

Control of purchased testing equipment, materials, and examination services shall include procedures to assure effective quality management.

Procedures must be established for selection and qualification of suppliers. Supplier surveys, past qualification history, periodic audits and evaluations, and, where appropriate, past performance may be used to establish qualification. A formal rating system may aid in determining the degree of control required by the purchaser, consistent with the complexity and quantity of purchase.

Receiving inspection should be conducted to the degree and extent necessary to determine acceptability. Receiving inspection should include historical records so that past supplier performance is available. Adequate facilities and procedures for maintaining separation of approved materials from unacceptable materials are necessary to ensure that nonconforming materials are not used.

A calibration system shall be established so that measuring and test devices can be calibrated, adjusted, repaired, or replaced before becoming inaccurate. Measurement, test, and inspection equipment shall be proven accurate by verification or calibration against certified standards before issuance for use. Periodic verification or calibration interval against the same certified standard shall be defined. When an item is found out of calibration, an evaluation procedure should be used to determine the validity of previous test or examination results.

All work instructions, examination procedures, specifications, and drawings shall be controlled for correctness and adequacy. The system shall ensure that correct revisions of applicable documents are available for use at locations where activities affecting quality are performed. Also, the system shall provide for the timely recall of obsolete documents.

The agency shall have a system to ensure that repetitive conditions adverse to the quality of the agency's work are identified and corrected. The corrective action program should be extended to suppliers, as appropriate.

Handling, storage, and shipping activities shall be covered by the quality system. Considerations should include handling damage, corrosion or infestation, degradation, loss from vandalism, and loss or obliteration of identifying markings. Periodic audits of stored items should be utilized to ensure against deterioration or expiration of shelf life.

As specified in ASTM E 543, the internal process forms or report of test or examination results should be filed with a written report to the client to allow traceability and become a part of the permanent record. The report should be accurate, legible, and pertinent. It should include the order number, specification, examination procedure, part identification, customer, and results of the examination. The reports should be signed by the technician performing the work and by a Level II or Level III individual. A procedure for Level III auditing of reports should be included.

E 1359

ASTM Guide for Surveying Nondestructive Testing Agencies (E 1359) establishes areas for review during the audit or survey and provides a uniform format that can be used in developing the information necessary for determining the technical competence of nondestructive examination agencies. ASTM E 1359 is divided into five parts dealing with facilities, system, equipment calibration, personnel, and the survey report or corrective action request. Some modification may be necessary to tailor the checklist to applicable areas and specific requirements.

Use of the survey form will aid in determining whether an agency has the capacity and capability necessary to meet the examination requirements. A review of the agency's policies and records will help to determine whether proper controls are in place and followed to assure proper implementation of requirements for nondestructive examination. Use of the survey form gives the auditor a permanent record and includes a corrective action request form.

Effective evaluations require technically competent evaluators or auditors. Laboratory surveys conducted by evaluators who are not technically competent cannot provide meaningful evaluations of qualifications for specific applications. Evaluators who are not technically knowledgeable can only perform a paper oriented evaluation. They can decide that an item exists but cannot evaluate the technical qualification adequately for particular applications.

Concluding Remarks

Consensus standards exist that provide guidelines for qualification, development of laboratory quality manuals, surveying, and auditing of nondestructive testing laboratories. Increased use of these standards by industry will result in generally higher standards for quality and reliability and should improve industry's position for work with the international community. Industry will be better served by assuring the qualifications of nondestructive testing laboratories before awarding contracts. Technically competent evaluations of NDT laboratories will serve to improve consistency in selection of qualified NDT services. Improved performance will result in minimizing schedule impacts and reducing costs.

Proper attention to evaluations of nondestructive testing agencies also will have the effect of equalizing the qualifications of contract bidders. Once a contract is awarded, it becomes very difficult and costly to replace the agency. Surveys or evaluations of nondestructive testing agencies should be conducted before the award of a contract or purchase order. Surveys conducted after an award of work may identify deficiencies, but then the activity becomes primarily one of corrective action to get the nondestructive examination agency qualified. When qualifications of organizations have been established in advance, awarding a contract to the lowest bidder does not impact on performance and the quality of results.

Effective evaluations require technically competent evaluators or auditors. Surveys conducted using a survey checklist developed from the E 1359 Guide with competent evaluators will provide meaningful evaluations of qualifications for specific applications.

Laboratory (agency) accreditation activity seems to be increasing in both the commercial and the government sectors and should serve to increase the confidence of the world community. The United States Department of Commerce's "National Voluntary Laboratory Accreditation Program" (NVLAP) is providing third party accreditation to assure an unbiased assessment of minimum qualifications for nondestructive testing contracts. Several governmental agencies, including the Navy, are requiring NVLAP accreditation. At least one independent nonprofit commercial organization, the American Association for Laboratory Accreditation (A2LA), is offering third party accreditation. As international activity increases, more and more emphasis on accreditation can be expected. The international community is very concerned about qualifications and will require independent third party accreditation to assure confidence in the test results.

Calvin W. Mckee[1]

Reverse Expansion—The Unification of Eight ASTM E-7 Glossaries*

REFERENCE: Mckee, C. W., **"Reverse Expansion—The Unification of Eight ASTM E-7 Glossaries,"** *Nondestructive Testing Standards—Present and Future, ASTM STP 1151,* H. Berger and L. Mordfin, Eds., American Society for Testing and Materials, Philadelphia, 1992, pp. 82–86.

ABSTRACT: ASTM Committee E-7 on Nondestructive Testing is composed of ten technical subcommittees. Volume 03.03 of the *Annual Book of ASTM Standards* presently contains eight glossaries which have been issued by these subcommittees, and other definitions can be found in an additional ten methods and practices. A review of these 17 documents has revealed several terms with two dissimilar definitions, and in some cases these definitions had been issued by one subcommittee. In an attempt to bring some order out of this chaos, we are proceeding as follows:

1. All terms (eight to date) which are common to the various disciplines have been placed under the jurisdiction of the editorial subcommittee.
2. A master glossary, which will contain all the terms currently found in the 17 documents, is being developed.
3. This glossary will be divided into sections, one for each subcommittee, and each of the technical subcommittees will retain jurisdiction over all the terms in its section. The eight common terms and any subsequent similar terms will appear in a separate section.
4. Where two or more definitions for one term currently exist, both will be shown until the subcommittee decides to eliminate one, combine them, or redefine the term.
5. A list of all terms in all the sections, showing in which section they may be found, will be appended to the master glossary.
6. If new subcommittees are added or new glossaries required, the terms will be added to the master glossary as a new separate section.

KEY WORDS: terminology, nondestructive testing, definitions, master standard

As all readers of *Standardization News* have been advised through the "Terminology Updates" by Wayne Ellis and Everett Shuman, the Committee on Terminology in its newly revised, and soon to be published, Part E of the *Form and Style for ASTM Standards,* more commonly known as the "Blue Book," will require that each of the ASTM technical committees "shall publish and maintain a general standard that contains all terminology published in all standards under the jurisdiction of the committee (including terminology standards in specific topics)."

Before getting into the approach for compliance with this requirement, I should like to present some idea of the problems that an extremely diversified committee, such as E-7 on Nondestructive Testing, encounters in attempting such a task. This will start with a brief discussion of the variety of techniques and methods which are included in this subject.

[1] Consultant, Engineering Materials & Processes, Inc., Wayne, PA 19087.

* Reprinted from *Standardization of Technical Terminology: Principles and Practices, ASTM STP 991,* 1988.

A short time ago one of the members noticed that none of the committee documents included a definition of *nondestructive testing*. We wrote one which has been approved. It is "The development and application of technical methods to examine a material or component in a manner that does not impair its future usefulness or serviceability and is performed to detect, locate, and measure discontinuities, defects and other imperfections; to assess integrity, properties and composition; and to measure geometrical characteristics."

Even if one isn't quite sure of a precise and satisfactory definition, nondestructive testing does play a major role in our lives. Medical doctors are among the foremost users of the art, and everyone has been examined by stethoscopes and sphygmometers to have the heart, lungs, and blood pressure checked. While tests of this type are thankfully not included in E-7's scope, anyone who has traveled by airplane has been subjected to two tests that are included. Hand-carried luggage is X-rayed, and one's person is checked by eddy current. The X-ray equipment is reportedly now being refined to the point where it will detect everything in the luggage but will not overexpose any film that might be there. The eddy current equipment is theoretically calibrated to let your change and keys pass, but not your guns and knives. The range of non-destructive test techniques is thus very large. This fact is represented both in the structure of Committee E-7 and, consequently, in its terminology.

At the present time, the committee has ten technical subcommittees and over 110 published ASTM standards, which range in designation from E 94 to E 1419. Each of these required a review to determine whether it contained definitions. The diversity of subjects did not make the work easier since the Committee grew somewhat like Topsy in *Uncle Tom's Cabin*. As new methods were introduced to the world, a new subcommittee was established, its only connection with its predecessors being that its activity was nondestructive testing in nature. The methods currently under the jurisdiction of Committee E-7 are (and the following are the basic definitions found in the subcommittee standards):

"Acoustic Emission is the class of phenomena whereby transient elastic waves are generated by the rapid release of energy from localized sources within a material, or the transient waves so generated." This subcommittee had a definition standard and three other documents containing definitions.

"Electromagnetic Testing, or eddy current, is a nondestructive test method for materials, including magnetic materials, that uses electromagnetic energy having frequencies less than those of visible light to yield information regarding the quality of the tested material." This subcommittee had a definition standard with additional definitions in three other documents. Twelve of the terms were defined in both the definition standard and one other standard and in no case were the two definitions exactly the same.

"Gamma and X-radiography are accomplished by the generation of a visible image produced by penetrating radiation passing through the material being tested." This group has a definition standard.

"Leak Testing comprises those procedures for detecting, locating or measuring, or combinations thereof, leakage." This subcommittee had a definition standard.

"Liquid Penetrant Examination is a nondestructive test that uses suitable liquids to penetrate discontinuities open to the surface of solid materials and, after appropriate treatment, indicates the presence of the discontinuities." It had a definition standard.

"Magnetic Particle Inspection is a nondestructive test method utilizing magnetic leakage fields and suitable indicating materials to disclose surface and near-surface indications." One definition standard.

"Neutron radiography is a process of making an image of the internal details of an object by the selective attenuation of a neutron beam by the object." This subcommittee did not have a separate glossary but had 15 definitions in one standard and an appendix labeled as a glossary in another standard. Two terms appeared in both places with dissimilar definitions in each case.

"Ultrasonic Examination is a nondestructive method of examining materials by introducing sound waves above the audible range into, through or onto the surface of the examination article." Definitions covering this subject were found in a definition standard and four other standards with the usual different definitions.

In addition, there is a subcommittee entitled "Other Methods," from which a glossary on infrared thermography had been ballotted and approved .

It can be seen from these descriptions that the methods have very little in common, and, while many use electricity to get started, once the working end of the wire is reached, all similarity ends. The total number of terms in all these standards is over 850, with a range of 160 definitions in one to two standards with only one definition included. Because of the differences in procedure, equipment, and vocabulary among the methods, there is little duplication of terms.

E07.92, the editorial subcommittee of Committee E-7, had been discussing ways of combining the definition standards issued by the various technical subcommittees for over a year, so it was not totally unprepared when the Terminology Committee requirement arrived. It had reviewed the printed documents and determined which ones had definitions. The 19 range from E 127 to E 1067 with 866 pages of Volume 03.03 between the first and the last, with the infrared thermography glossary not yet included. While doing this research, we could not help realizing that Committee E-4 on Metallography, with whom E-7 shared the volume, had only two glossary standards, one on metallography and one on heat treatment of metals. These appeared sequentially in the book. Despite the many types of materials which are examined metallographically, the similarities in procedure have enabled E-4 to hold the number of their glossaries to a minimum. In the review several terms with dissimilar definitions were found, and in 22 cases these had been placed in different documents by the same subcommittee. Each subcommittee chairman was given a list of the duplications found in their standards in the hope they would attempt to resolve the situation. It was also found that the same word may mean different things to different subcommittees all within the nondestructive testing committee scope. For example, "develop" in radiography is what one does with film to bring out the image, but in liquid penetrant examination, it is the application of a liquid suspension to the test area to draw penetrant from discontinuities.

The spread from E 127 to E 1067 was guaranteed to become greater as time goes on, simply because of the ASTM system of assigning the next sequential number to the next standard which fits in a category. This situation is similar to the universe, which is regarded by astronomers as a constantly expanding operation. Accordingly, one of the objectives is what might reasonably be called "reverse expansion." Thus the compilation of the "general standard" had two basic aims, first to meet the Committee on Terminology requirements, and second to get from multitude to solitude. This phrase has been borrowed from C. Northcote Parkinson and, even though he used it to describe the shift of a bride from the typing pool to the kitchen sink, it aptly portrays the situation.

In most of the companies in which E-7 members are employed, despite the tremendous differences in technique, there is usually only a single section entitled "Quality Assurance," or its equivalent, with both nondestructive testing and inspection included in it. As new methods are introduced, the nondestructive testing personnel are assigned to learn all about it, since of course to management "Nondestructive testing is nondestructive testing, isn't it?" Accordingly, many of the members serve on more than one subcommittee. Since this is the case, an individual may have to deal with more than one method in a fairly short time, and no one should be required to look through 866 pages to find a definition.

The first idea in combining the terms in all the standards was a simple alphabetically arranged dictionary-type listing. However, further thought indicated that this would probably result in total confusion to someone who was endeavoring to look up words in a single method. It was also realized that despite having representation and input from all of the subcommittees, the editorial subcommittee was not capable of assuming jurisdiction over all the E-7 terms. Furthermore, and very correctly, the technical subcommittees would not have yielded their rights and privileges in this field.

Accordingly, the final concept of how best to comply with the Committee on Terminology directive, still permit the technical subcommittees to retain control of their own vocabularies, and prevent the further dispersion of nondestructive testing terms by the addition of future definition standards (which would have numbers higher than E 1067) is the following three-step procedure:

First—Terms which are common to several subcommittees have been taken from their respective standards and, along with the final accepted definition of nondestructive testing, are placed under the jurisdiction of E07.92. Most words of this type have to do with what one is looking for, such as defect, discontinuity, and indication, and variations on these, such as false indication or nonrelevant indication. Many of these words appeared in several of the subcommittee standards with minor differences in phrasing. In fact it was the proliferation of definitions of the word "defect" (attributed to E-7 five times in the ASTM compilation) that was the original inspiration to develop a master standard. The placing of these common words under the jurisdiction of E07.92 was considered the best approach to reduce the number of definitions per word to one.

Each of the technical subcommittees has its editorial chairman automatically included as a member of E07.92, so that all of them have an opportunity to contribute to the definition of the common terms. E07.92 has just completed a ballot of "defect," and all future additions to this section will be ballotted in E07.92 rather than one of the technical subcommittees prior to the E-7 ballot.

Second—The Master Glossary or "single standard" is not a single alphabetical listing, but is a classified listing divided into sections, one for each of the various subcommittees. Every technical subcommittee retains full jurisdiction over its section. The E07.92 common terms appear in a separate section. Where two definitions for one term currently existed in one subcommittee's standards, both are included. The number of the standard, other than the current glossary, is shown with the second definition. The intention is to include both definitions until the subcommittee decides to eliminate one or combine the two by redefining one of them.

Third—An alphabetical list of all the terms included in the master glossary, showing in which section each may be found, is appended.

As new standards, such as the one on infrared thermography, are added to the committee documents, their terms will be given a new separate section in the master glossary, with the words also added to the alphabetical list. New words in the existing standards will be added to both the proper section and the list. Definitions in new standards will be treated similarly.

It was decided at a meeting of Committee E-7 that the new combined glossary should appear in the gray pages of the *Book of Standards* as a proposal document for two years, which gave everyone a chance to eliminate any bugs that emerged. It gave also all of the members of the Committee an opportunity to use the proposal standard, make any suggestions for improvement, and find that when it was officially issued it would not be a surprise. It gave E07.92 a chance to see how well it could handle any ballotted definition changes to keep the standard up to date, demonstrate that control of the master standard was possible, and with this control overcome the technical subcommittees' reluctance to change their procedures. E07.92 had a key role in updating the master standard during this interim period, since changes and additions to the present standards and definitions included in new standards do not automatically appear in the proposal document. However, since all changes and new standards require both a subcommittee and committee ballot, it is expected the first will alert proposed changes and the second will tell what the actual changes are. This way, when the subcommittee standard is published, the proposal standard should be current.

Fortunately, once the master standard was approved, changes in definition standards were ballotted as changes in the subcommittee section, and E07.92 was then mainly concerned with definitions in the method standards. In this, E07.92 has a management position, but the subcommittees maintain their technical jurisdiction. No major conflict was anticipated in these roles.

At the end of the two-year trial period, each of the technical subcommittees was asked to vote to transfer its glossary from the present individual location to the master glossary section assigned to it. Upon completion of this vote the new combined glossary became the official document for the definitions of Committee E-7 and the individual definition standards were withdrawn. The terms presently found in the methods standards continue to be found in those standards, unless the subcommittees decide to remove them, this action being their responsibility.

This arrangement provides a satisfactory solution to the Blue Book Part E requirement, prevents the future spread of definitions through the *Book of Standards,* and hopefully provides a model to other Committees which are faced with a similar problem.

While it is realized that the end result of this endeavor puts the new master standard near the back of our volume of the *Book of Standards,* simply because it was given the next E standard number upon adoption, at least all the definitions are in one place.

In taking care of the needs of Committee E-7, as well as complying with the Committee on Terminology requirement for a master standard, it is very evident that neither of these objectives would have been possible without an editorial subcommittee to perform the actual compilation. None of the technical subcommittees had the expertise, and the Executive Committee was not the place for work of this type. However, by assigning the task to the editorial group, the Executive Committee performed its obligation, and, through their representatives on the editorial subcommittee, the technical subcommittees made their contribution to the final product. Without editorial subcommittee direction, the master standard would still be in the talking stage.

NDT Standards: NIST, DoD, ASME, SAE, ISO, EC

George Birnbaum,[1] *Donald G. Eitzen,*[1] *and Leonard Mordfin*[1]

Recent Developments in NDE Measurements and Standards at NIST

REFERENCE: Birnbaum, G., Eitzen, D. G., and Mordfin, L., **"Recent Developments in NDE Measurements and Standards at NIST,"** *Nondestructive Testing Standards—Present and Future, ASTM STP 1151,* H. Berger and L. Mordfin, Eds., American Society for Testing and Materials, Philadelphia, 1992, pp. 89–125.

ABSTRACT: This review discusses the NDE measurements and standards developed at the National Institute of Standards and Technology (NIST) during the last decade. These measurements and standards, which have been primarily developed to meet the needs of postmanufacturing and in-service inspection, include those for ultrasonic and acoustic emission methods; electromagnetic methods (eddy currents, magnetic measurements, and light scattering and capacitive techniques for surface roughness); and miscellaneous methods (infrared thermography, leak testing, visual acuity, and X-ray radiography). The theoretical and experimental bases of these methods are emphasized wherever possible.

KEY WORDS: ultrasonics, acoustic emission, eddy currents, magnetic methods, optical scattering, infrared thermography, leak testing, visual acuity, X-ray radiology

In the application of nondestructive measurements to determine whether materials and structures are suitable for their intended use, it is necessary to ascertain the reliability and accuracy of these measurements. The performance of nondestructive evaluation (NDE) equipment and the validity of the ensuing measurements are determined by utilizing a variety of standards. Practically, standards assume a variety of forms including: artifacts with theoretically known or calibrated defects; standard reference materials (SRMs); methods for determining the characteristics of NDE probes; comparison of NDE measurement systems with measurement systems of known accuracy, i.e., calibration procedures; and documentary standards. The last are the means of archiving and disseminating the rules for measuring a given quantity.

Our purpose is to present an overview of advances in all kinds of NDE standards developed at NIST during roughly the last decade. These standards have been developed primarily to meet the needs of postmanufacturing and in-service inspection. Wherever possible we tend to emphasize the theoretical and experimental bases of the standards. The large number of methods that we discuss are organized according to the following outline. Some new work not previously published is presented in the sections on Acoustic Methods, Magnetic Methods, and Optical Scattering.

Acoustic Methods
Primary Calibration Facility for Acoustic Emission Sensors
Tools for Secondary Calibration of AE Sensors

[1] Senior scientist, Office of Intelligent Processing of Materials; Group Leader, Ultrasonic Standards; and Group Leader, Mechanical Properties and Performance, respectively, National Institute of Standards and Technology (NIST), Gaithersburg, MD 20899.

Ultrasonic Reference Blocks
Ultrasonic Radiation Force Balance
Ultrasonic Applications Standards
Electromagnetic Methods
Eddy Current Measurements and Standards
   Eddy Current Equipment Calibration on a Fundamental Basis
   Probe Characterization
   Artifact Standards
   Eddy Current Probe Sensitivity as a Function of Coil Construction Parameters
   Uniform Field Eddy Current Probe
Area Averaging Techniques for Surface Roughness
   Optical Scattering
   Parallel Plate Capacitance Gauge
Magnetic Methods
Magnetic Particle Inspection
Barkhausen Noise
Miscellaneous Methods
Infrared Thermography
Leak Testing
Visual Acuity
X-Ray Radiology
Concluding Remarks

Acoustic Methods

Following the terminology of ISO Committee TC 135 on Nondestructive Testing, acoustic methods cover both acoustic emission and ultrasonic testing. Both of these techniques are largely comparative (as opposed to absolute) and rely on the characteristics of the acoustic source, modification of the signal by the material under test, the characteristics of the receiving transducer, and the processing and interpretation of the received signal. These attributes of acoustic methods place significant demands on components and system calibration and standardization, e.g., transducer characterization and calibration. NIST (formerly NBS) has been actively developing the theoretical and experimental bases, tools, and facilities for the calibration and characterization of devices and quantities for acoustic methods. When appropriate, these developments have been codified and promulgated as documentary standards; however, in all cases, an account of these developments has been published [1a].

Primary Calibration Facility for Acoustic Emission Sensors

An acoustic emission (AE) sensor calibration facility was developed at NIST [1] and is being used to calibrate sensors and secondary transfer calibration transducers. The facility is also being used to invent and develop new transducer technology for AE applications and standardization [2,3].

An early and fundamental question to resolve regarding a new calibration scheme has been, "What are the physical quantities to be measured and under what conditions?" One sensor characteristic of interest is its sensitivity, i.e., its output per unit input. The practical output quantity for all AE sensors is a voltage-time waveform or a quantity derived directly from it. The appropriate input quantity to be measured is not so obvious. The input to the sensor may be considered to be the motion at the location where the sensor is mounted *or* the stress (surface traction, pressure, force) at the surface of the sensor. The magnitude of this stress, however, is strongly dependent on the characteristics of the sensor. For example, the boundary stress for an optical probe is sensibly zero. Thus, basing a definition of sensitivity on stress is erroneous. Consider, for example, two sensors with different acoustic impedances mounted on a test block symmetrically with respect to an AE source. Assume for the same event that the voltage outputs of the two sensors are the same. Their calibrated sensitivities would be different because their stress inputs were different even though the AE source was the same and the outputs were the same.

The motion at the mounting-surface interface is also affected by the presence of a sensor, but far less than the stress. This difficulty is avoided by considering, as the input for the calibration, the motion of the surface in the absence of the transducer. However, this will not work for the normal stress because in the absence of the transducer this stress is zero.

Thus, we define the calibration of AE transducers in terms of the voltage output per unit of motional input which would occur without the transducer present. This definition is a very practical one since normally the AE event and signal in the solid are of primary interest, not the signal modified by interactions between the sensor and the solid. There remains the problem of how to determine this motional input.

The interactions between a sensor and the solid on which it is mounted depend on the acoustic impedances of both. Thus, the calibration of a sensor will, in general, be different when coupled to different materials. The material chosen for the primary calibration block was ASTM A 508 class 2 steel partly because it is very similar to many pressure vessel materials. The difference in the measured sensitivity of a sensor on different materials depends on the properties of the sensor, so general statements are not possible. However, for one transducer, which was carefully analyzed [4] and calibrated on several materials, the results show that the sensitivity of this transducer differed by only 10% mounted on aluminum as compared with steel but by 95% mounted on PMMA as compared with steel [5]. This implies a strong need for a calibration system for composite materials.

The motional input (i.e., the motion at the location on the mounting surface where the transducer is to be placed but with it absent) for the AE transducer calibration is determined two ways: elasticity theory and measurement with a standard transducer. During the design of the calibration scheme, the most complex solid shape for which the exact transient elastic solution was known was that for an infinite half space subject to a normal point-force step function on the surface. The input to the theory is the magnitude of the step function and the speeds of sound in the material. This solution gives the displacement as a function of time for any point on the surface. This waveform is termed the "seismic surface pulse" and is shown in Fig. 1.

The assumptions of this theoretical solution dictated all important aspects of the AE calibration facility; that is, the calibration scheme was designed to model the theory in every practical way. While the structure on which the transducer is mounted is not an infinite half space, it is large enough (a cylinder about 1 m in diameter and 1/2 m thick) so that within the assumptions of linear elasticity it behaves exactly like an infinite half space for about 120 μs after the start of the step-function force input. For well-behaved transducers, this time is long enough to obtain the necessary calibration data and maintain correspondence with the exact theory.

In addition to theory, the motional input to the transducer under test (TUT) is determined by measuring it with a standard absolute dynamic displacement transducer. This standard transducer was designed and constructed so that its absolute sensitivity is known by calculation [6]. The displacements determined by theory and measured by the standard transducer agree within a few percent. During a calibration, the standard transducer and the TUT are both placed on the same surface of the block, equidistant and in opposite directions from the central location of the point source. A schematic of the calibration apparatus is shown in Fig. 2. This symmetry of positioning assures that the motions at the locations of the standard transducer and at the TUT are identical. Comparison of the output of the TUT with the motional input

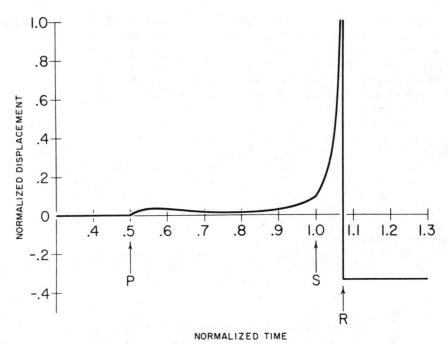

FIG. 1—*Theoretical seismic surface pulse. P, S, and R are the longitudinal, shear, and Rayleigh wave arrivals, respectively (Ref 1).*

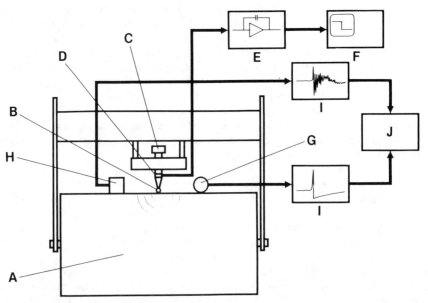

FIG. 2—*Schematic diagram of the calibration apparatus for calibrating acoustic emission sensors. A = steel transfer block, B = capillary source, C = loading screw, D = PZT disc, E = charge amplifier, F = storage oscilloscope, G = standard transducer, H = transducer under test, I = transient recorders, J = computer (Ref 1).*

measured by the standard transducer (or calculated from the theory) yields a calibration of the TUT.

A calibration geometry slightly different from that of Fig. 2 is sometimes used. In this geometry the point source is again at the center of one face of a cylindrical calibration block and the transducer under test is at the center of the opposite face of the block. Comparison of the TUT output with the calculated motion at the location of the TUT gives a calibration of the TUT. This calibration is referred to as the through-pulse method. It does not include measurement of the TUT's aperture effect [5], which is important in the presence of surface waves.

A fundamental aspect of the calibration is the experimental technique for generating a point-force step function input. The technique is described more fully in Ref 1. The step-function force events are generated by breaking a glass capillary on the surface of the calibration block. The capillaries are drawn at high temperature from laboratory borosilicate glass tubing. They have a nominal outside diameter of 0.2 mm and a nominal inside diameter of one third of this. The capillary is placed on and parallel to the surface of the block at its center. A solid glass rod 2 mm in diameter is placed on and perpendicular to the capillary and parallel to the surface of the block. The rod is forced toward the capillary and the block surface by a loading screw [1]. By slowly turning the loading screw, the load on the rod, capillary, and block surface is quasistatically increased until the capillary breaks and the force on the block goes to zero and stays at zero. This unloading occurs in about 0.1 μs. The loading screw contains a piezoelectric disc which is calibrated to measure the magnitude of the step-function force, which is typically about 20 N. The force differs from event to event but is measured so that the motional input to the TUT can be calculated.

The force-time history of the breaking glass capillary event has recently been independently measured with unprecedented accuracy of a few percent [7]. A special capacitance receiver was designed and constructed and a special test block prepared so that the epicenter transient displacement due to seven different types of force events could be accurately measured. The seven types of force events ranged from ball impacts to breaking capillaries to small quantities of high explosive. The measured displacements were convolved with the inverse Green's function and the inverse of the impulse response function of the receiving transducer and electronics. The result is the remotely measured source force-time waveform. The result for the breaking glass capillary event is shown in Fig. 3, which shows the waveform to be a very good approximation to a step function and shows the spectrum to be rich in content to several megahertz.

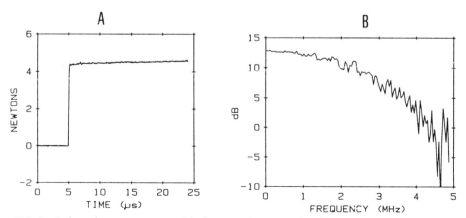

FIG. 3—*Independent measurement of the force-time history produced by a breaking glass capillary:* (A) *waveform from 0.117-mm capillary,* (B) *spectrum of the waveform (Ref 7).*

The calibration process indicated in Fig. 2 can be summarized as follows (details are given in Ref *1*). The TUT and standard transducer are mounted symmetrically with respect to the capillary source on the large (2200-kg) steel block. The capillary is compressed against the block by the loading screw until it breaks, producing a step-function force the magnitude of which is measured by the piezoelectric disc. The loading effect of the TUT on the surface motion is part of its calibration; the standard transducer does not sensibly load the surface. The transient voltage outputs of the two transducers are digitized and stored for processing. Complex spectra are determined for the two transducers using a fast fourier transform. The calibrated response of the TUT is determined by dividing the spectrum of the TUT by the standard spectrum, frequency by frequency. The magnitude and phase response are derived from this quotient spectrum in the usual way. There are several alternative ways for expressing the calibration data. The most natural unit for the magnitude of the TUT sensitivity is volts of output per meter of surface displacement since the standard transducer is a displacement sensor. The response data can easily be converted to volts per unit velocity. The calibration data can also be presented as a time-domain impulse response.

This calibration scheme was used as a basis for an ASTM standard (ASTM E 1106–86), and a draft ISO standard on the primary calibration of AE sensors.

Tools For Secondary Calibration of AE Sensors

NIST has essentially completed the development of a scheme for the secondary calibration of AE sensors and is in the process of writing a draft standard for submission to ASTM. The expected users of the scheme are AE sensor manufacturers/suppliers, laboratories, and testing services. The emphasis here is on the tools developed at NIST which form the basis for secondary calibration schemes.

A theoretical tool is the exact transient elastic solution for wave propagation in an infinite plate. The theory was developed independently at NIST [8] and at Cornell University [9] by slightly different methods. Later, the computer code which performs the calculations was made available by NIST in a user friendly version which runs on a PC [10] and has since been used by many universities and laboratories. The theory gives the transient displacement at any selected near field point due to any point force-time waveform input. The inputs to the theory are the thickness and wavespeeds of the plate, the force-time waveform source, and the desired observation point. The correspondence between the theory and an experiment with a transient source on a finite plate is exact (within linear elastic assumptions) until the time when a wave from the source travels to an edge, reflects, and reaches the observation point. The theory thus provides a powerful tool for studying calibration sources, computing the motional input for a convenient calibration structure, and for calibration quality control.

An important experimental tool for the secondary calibration of AE sensors is the NBS conical transducer [2,11]. It consists of a truncated cone of PZT mounted to a large brass backing. The small end of the cone contacts the surface to be measured, creating a small aperture for the arrival of ultrasound. The transducer was designed to be very sensitive (about as sensitive as a resonant commercial AE sensor and much more sensitive than an interferometer), very broadband, and to have a constant sensitivity over its working frequency range as is illustrated in Fig. 4. This transducer measurably detects only dynamic motion normal to the mounting surface. A principal reason for its development is to serve as a transfer standard for the secondary calibration of AE sensors, and it is available from NIST as Standard Reference Material 1856.

A transducer was also developed to accurately measure motion tangential to the mounting surface [12], but its characteristics have not been as fully exploited. An example of its behavior is shown in Fig. 5, in which a theoretical tangential displacement waveform is compared with the waveform measured with a NIST tangential transducer.

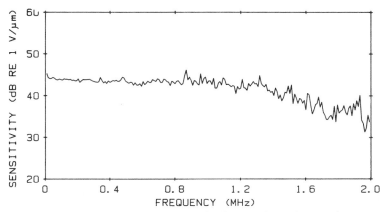

FIG. 4—*Sensitivity versus frequency for the conical transducer (Ref 2).*

Some of the utility of the first two tools is demonstrated by Fig. 6, which compares theory and experiment for the dynamic normal displacement of a plate. In the experiment a point-force step function was generated on the surface of a steel plate by breaking a glass capillary. An NBS conical transducer was placed on the same surface of the plate about five plate thicknesses away from the source to measure the resulting waveform. A theoretical waveform was calculated using a step-function input and the measured wave speeds of the plate. The two waveforms, measured and calculated, are shown in Fig. 6; they are offset for clarity. The fine structure in the waveforms is due to multiple reflections and mode conversions at the surfaces of the plate.

Results such as these confirm that dynamic elasticity theory and, specifically, the exact code contained in Ref 10 provide a useful and accurate tool for describing AE wave propagation in specific engineering structures and for predicting the motional input displacement for secondary sensor calibration. The results also confirm the NBS conical transducer as a valuable tool for measuring dynamic displacements and as a transfer standard for sensor calibration.

Yet another tool is the breaking pencil lead source (designated by ASTM as the Hsu source), which has been a useful calibration tool for some time [13]. It continues to be more widely used nationally and internationally. The technique and hardware continue to be additionally controlled, standardized, and codified [14] for calibrations and checks of AE systems and sensors. Recently the force-time history generated by the Hsu source has also been measured with unprecedented accuracy [7]. The procedure is the same as that mentioned above for the determination of the capillary waveform. The result is shown in Fig. 7. The negative precursor to the step function is real and can be explained by the details of the wave propagation at the time of the lead break. The rise time of the step is not as rapid as with the glass break. The time signal of the glass break does not have the negative precursor, but the convenience and lower skill requirements of the lead break make it the technique of choice in many standards applications.

Ultrasonic Reference Blocks

NIST/NBS has operated a calibration service for ASTM E 127-type [15] ultrasonic reference blocks since 1976. These artifacts are right circular metallic cylinders with a hole drilled 0.750 in. deep on the axis of the cylinder and terminated with a flat bottom (see Fig. 8). Although the ASTM E 127-type block is made of 7075 aluminum alloy, the block may be made of other materials such as steel and titanium according to another ASTM standard.

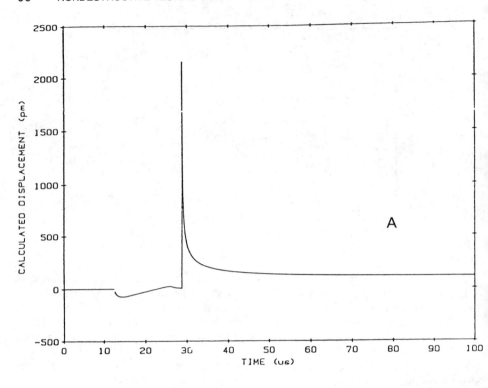

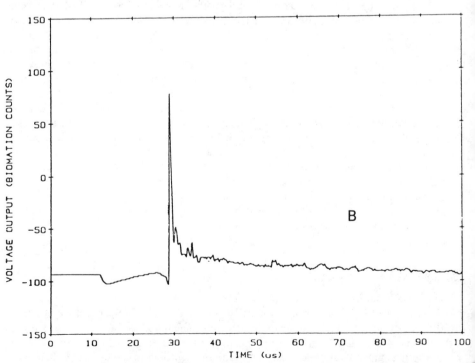

FIG. 5—*Theoretical* (A) *and experimental* (B) *displacement waveform measured with the tangential transducer, produced by breaking a glass capillary (Ref 3).*

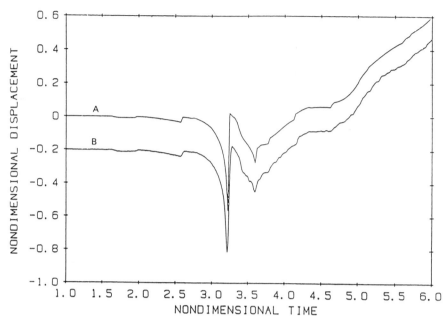

FIG. 6—*Comparison of displacement from exact theory* (A) *and measurement* (B) *on a plate resulting from a point-force step-function input. The time window is 60 μs; the displacement is about 10 nm. The waveforms are offset for clearer comparison.*

These blocks have various lengths between the top surface and the hole termination and various hole diameters. In use, an ultrasonic pulse enters the top surface of the block, travels along the axis of the block, and strikes the flat bottom of the hole. Some of the energy is reflected back to the ultrasonic transducer, which also acts as a receiver. This reflected energy is used as a reference signal for many functions in ultrasonic testing including the setting of accept/reject

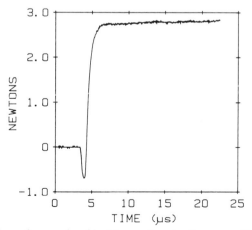

FIG. 7—*Measured waveform produced by 0.5-mm-diameter Hsu source (JSNDI-006 pencil and lead) (Ref 7).*

Ultrasonic Wave Entry Surface

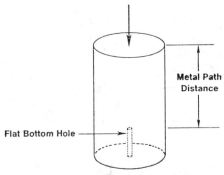

FIG. 8—*Schematic of an ultrasonic flat-bottom hole reference block.*

criteria for raw stock and parts. For this reason it is essential that the artifacts produce a specified ultrasonic amplitude response.

One of the customers of this calibration service brought to our attention an extremely unusual set of blocks and permitted us to investigate this "rogue" set nondestructively and destructively [17]. The ultrasonic response of this set, shown in Fig. 9a, is 600% larger than from the expected response for the longest blocks, which is both alarming and surprising. Further, the response is high by that much when most of the variables for blocks tend to lower their response, e.g., higher attenuation, nonparallel surfaces, rounded corner on the hole. Ultrasound revealed no other unusual characteristics; x-rays revealed no unusual geometry; metallography revealed nothing unusual about the material.

More recently one of the blocks was sectioned to obtain a solid disc of material. This disc was further analyzed ultrasonically and compared with a disc of material from a block exhibiting expected ultrasonic response from its flat bottom hole. These discs were insonified by 5 MHz ultrasound while immersed in water. The ultrasound was transmitted from the transducer, through the water, through the disc, and into the water beyond the disc. The sound field which had propagated through the disc was probed using a broadband hydrophone to map the relative spatial pressure field. The sound field after having propagated through a disc from a "rogue" block and a disc from a normal block are shown in Fig. 9b. It is clear from the pressure map that the disc from the "rogue" block focused the ultrasound onto the axis of the block. As more of this material was traversed by ultrasound, the pressure field became higher and higher where the flat-bottomed hole was located. Thus, for short blocks their response was near normal, but for longer "rogue" blocks the response became very high as shown in Fig. 9a.

Ultrasonic Radiation Force Balance

The NIST ultrasonic radiation force balance has been a calibration resource for some time. The apparatus [16] measures the total absolute ultrasound power output of a transducer radiating into water while the transducer is being driven at a carrier frequency which is variable over a wide range, 0.1 to 50 MHz, and is modulated at a low frequency so that the detector acts as a narrow-band receiver. The ultrasonic power output is equal to the measured radiation force divided by the speed of sound in water. The radiation force is sensed by one coil magnet and is opposed and nulled or balanced by a second coil magnet. This radiation force is equal to the $B\ell$ product of the coil magnet (where B is the magnetic field strength and ℓ is the number of turns in the coil) times the current required to balance the force.

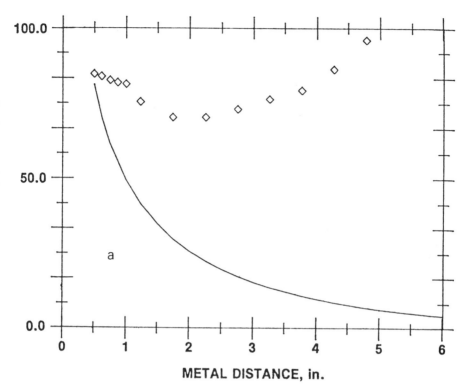

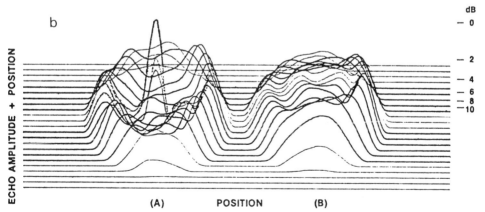

FIG. 9—(a) *Echo amplitude versus metal path distance from a "rogue" block (diamonds) and a typical date set (solid line). (b) Back surface echo amplitude versus position for two like-geometry solid aluminum cylinders:* (A) *a block from the "rogue" set with the hole removed; and* (B) *a cylinder of normal bar stock (Ref 17).*

The $B\ell$ product can be measured with d-c current using analytical weights to determine the force and from the value of the current. Thus the power output of an ultrasonic transducer can be calibrated in an absolute way. The development of this apparatus was motivated by a desire to monitor the condition of NDE transducers by obtaining their power versus frequency curves. In fact, the most frequent use of the apparatus is for ultrasonic medical dosimetry and for the calibration of standard ultrasonic power sources.

The radiation force balance suffered from certain limitations. The upper suspension assembly of the balance developed over time a measurable hysteresis which, although small, was noticeable at the highest sensitivity levels (least count sensitivity of about 10 μW). The procedure for measuring the $B\ell$ product of the balancing coil was tedious and time consuming, requiring hours of setup and stabilization; however, the stability of this product is a fundamental measure of the stability of the device. Furthermore, the electronics of the force balance required sinusoidal modulation of the transducer-drive voltage as well as sinusoidal a-c current to generate the balancing force. These requirements limited the application of the force balance to transducers driven by sinusoidal modulation in contrast to the excitation used in most ultrasonic systems.

The radiation force balance was recently rebuilt (the balance itself and its ancillary electronics) to eliminate these limitations and to tighten the bounds on the largest uncertainties. The system was disassembled and the original upper suspension which developed hysteresis was replaced with a tensioned-wire suspension assembly which is immune from this effect. A dual-quadrature Michelson interferometer was designed, constructed, and installed in order to easily determine the absolute position of the target during redeterminations of the $B\ell$ product.

The electronics were completely redesigned to use rectangular-envelope modulation of the drive and nulling waveforms. This resulted in an improvement in the uncertainty of drive voltage and nulling current by up to a factor of five and the reduction of timing and phase shift errors to negligible levels. It also allows the balance to be used with complete ultrasonic systems to measure, in minutes, power levels that once required days using our previous methods.

Ultrasonic Hydrophone Calibration Development—Hydrophones, miniature underwater receivers, are used extensively by ultrasonic technologists to measure the transient pressure waveforms by ultrasonic transducers. The sensitivity of a hydrophone in terms of volts per unit pressure versus frequency is not currently traceable to U.S. national standards. The sensitivity of hydrophones is not known to a specified accuracy and can be unstable in certain cases. A calibration service for hydrophones has been frequently requested by researchers, regulators, equipment manufacturers, and associations supporting ultrasonic measurements and is a required measurement service in some documentary standards.

Motivated by this need, a project to develop a measurement service for characterizing hydrophones was recently begun. Several technical strategies were considered. An interferometric technique has demonstrated accuracy and independence from other methods but requires extensive resources. Various time-delay-spectroscopy schemes have advantages for high-volume testing but also require extensive resources. With the modest resources available for this project, three methods were analyzed: planar scanning, reciprocity, and substitution. Reciprocity methods applied to hydrophones were judged to result in unacceptably high uncertainties in the determination of sensitivity. Substitution methods require the existence and use of a hydrophone with an accurately known sensitivity and proven long-term stability for use as a national standard; however, the development of a hydrophone with such properties is problematical. The planar scanning technique was judged to be implementable with available resources including optimal use of the radiation force balance that was recently improved as outlined above.

The planar scanning technique is based on the use of a source transducer which projects a known total power (as accurately determined by the NIST radiation force balance), and the measurement of the voltage waveform output of a hydrophone. Measurements must be made at all locations in a scan plane perpendicular to and in the far field of the source transducer. The total power output is equal to the surface integral of the intensity distribution in the scan plane. But the intensity is proportional to the square of the pressure, which is proportional to the hydrophone output voltage. Thus by numerically integrating the square of the hydrophone output voltage in the scan plane in the presence of a known power output from the source, the sensitivity of the hydrophone can be determined at any frequency of interest.

The development of a calibration system began with the acquisition of a commercial turn-key system. Early testing revealed that the hardware was adequate but that the software was hopelessly flawed and the most efficient fix was a complete rewrite. As a result of the rewrite, the system was able to store and analyze up to 2400 waveforms per planar scan, each waveform consisting of 500 samples of eight-bit data. By recording the actual waveforms rather than extracted parameters, various algorithms for data reduction could be tested.

The hardware and software have been used to test the method by repeatedly calibrating a hydrophone with various calibrated sources operating between 1.6 and 10 MHz. Calibration data provided by the National Physical Laboratory (NPL) for this hydrophone in 1985 provides a basis for checking both the stability of the hydrophone and the accuracy of the technique currently under development. For frequencies up to 5 MHz, the agreement between the 1985 NPL and our results for values of pressure-voltage sensitivity seem quite good. For example, the average of twelve calibrations done at various distances using a NIST source operating at 1.6 MHz differed from the NPL calibration by only 5%, with the standard deviation being 3% of the mean value.

Plans for the immediate future include expanding the data base to a range from 1 to 20 MHz. This will require acquisition of additional smaller hydrophones and a different digital capture device. Some issues regarding signal processing (compensation for d-c offsets, subtraction of noise floor signals) still exist. Considerable work remains to establish useful uncertainties and to begin the new measurement service.

Ultrasonic Applications Standards

Recent projects on automated noncontact ultrasonic gaging have raised issues of standards for use during materials processing. A system was developed to measure the thickness of shells from one side while they are in a vacuum chuck at 60 000 locations to an accuracy of ± 2.5 μm in a matter of minutes. Initial setup required the use of artifact standards to align the ultrasonic axes with the machine tool axes. An on-machine thickness artifact made of the part material was also required to achieve the required accuracy [18].

In a different project on monitoring part area-average surface finish using ultrasound coupled through a stream of cutting/cooling fluid, an artifact standard was found to be needed. (Other techniques for monitoring average surface roughness are discussed below in more detail.) The artifact required must be of the same material as the part being monitored and must have a surface smooth compared with the part [19].

Further, high-frequency tightly focused ultrasound has been used to profile surface finish. The standards issues for this application have not been totally resolved, but specimens with a spatial sinusoid of various known amplitudes and wavelengths will almost certainly be involved [19].

Electromagnetic Methods

Eddy Current Measurements and Standards

Calibration of eddy current systems [20], as in other NDE measurement systems, is an important factor for attaining the accuracy and precision of measurement that quantitative nondestructive evaluation requires. The quantity of interest in most forms of eddy current inspection is ΔZ, the change in probe impedance induced by a flaw. Flaw signals, such as those produced by surface-breaking cracks, are small and may be easily obscured by the impedance changes caused by small variations in the height of the probe above the work piece (lift-off). Thus, in the determination of flaw signals, it is necessary to discriminate against lift-off. Several possibilities exist to calibrate eddy current equipment on a fundamental basis: (1) comparison of experimental lift-off data with theory; (2) comparison of theoretical and experimental probe response of flaws that can be analyzed accurately; and (3) insertion of small resistances in series with the probe to provide fiducial marks on the response obtained for unknown flaws [21].

Eddy Current Equipment Calibration on a Fundamental Basis—The magnitude and spatial distribution of the magnetic field of air core probes can be accurately calculated using the theory of Dodd and Deeds [22]. Measurement of lift-off signals obtained with an air core probe were found to be in good agreement with the lift-off response calculated from this theory. However, it is considerably more difficult to calculate probe impedance changes produced by flaws because of the greater geometrical complexity of a flaw compared with a smooth surface. Nevertheless, solutions for the response of an eddy current probe to flaws are now available for the cases of rectangular and semielliptical surface-breaking flaws interrogated by a nonuniform probe field [23,24]. Of course, it is necessary that the field distribution of the probe be known either by calculation or measurement [25], and that the measurement system used to determine changes in probe impedance be of known accuracy. The results of flaw-signal measurements at several frequencies for an EDM notch in Al6061 for the magnitude of the impedance change, ΔZ, plotted against the position of the probe relative to the center of the flaw, normalized by r, the flaw length, are shown in Fig. 10 together with the theoretical predictions. Actual differences between theory and experiment were approximately 25%, well within the

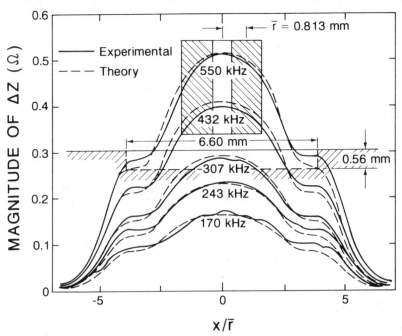

FIG. 10—*Magnitude of ΔZ determined by scanning the probe along the length of an EDM notch in Al 6061 alloy. Theoretical curves were calculated from a nonuniform-field probe-flaw interaction theory. The abscissa represents normalized probe position relative to the center of the flaw (Ref 21).*

experimental uncertainty. Closer agreement between theory and experiment was obtained in the lift-off measurements.

The insertion of calibrated small resistances in series with the probe was shown to be capable of calibrating the entire eddy current measurement system for impedance changes and may be used with either air core or ferrite core probes in either differential or absolute probe configurations. However, this measurement does not say anything about the probe sensitivity or provide a characterization of the probe in its interaction with flaws. Calibration methods for accomplishing this based on comparing measurements of probe response to theoretical predictions is of importance in establishing the method on a fundamental basis but requires a knowledge of the probe's magnetic field intensity and spatial distribution. This information can be calculated for specially constructed air core probes; however, little is known about the field of commercial air core probes where the lay-up of the windings is not precisely controlled. With ferrite core probes, the problem is much more complex and the theoretical results are uncertain. Thus it is helpful to be able to map the field profile of air core probes and particularly ferrite core probes.

A magnetic field map or profile, in addition to its need for the above studies, is a good tool for examining the variations in probe performance. Such a map was obtained by coupling a room temperature field mapping (or search) coil to the input coil of a superconducting quantum interference device (SQUID) in liquid helium [26]. This system is very sensitive and allowed the use of very small search coils. With this device, the fields of a number of air and ferrite coils were mapped in considerable detail. This work revealed that eddy current probes of nominally identical construction had pronounced differences in their magnetic field magnitudes and distribution.

Although this work is important in establishing eddy current measurements on a fundamental basis, it does not provide an answer to the practical problem of how to ascertain the reliability and accuracy of eddy current testing in such applications as conductivity measurements, metal sorting, checking heat treatment, and detecting fatigue cracks and other discontinuities. To deal with these practical problems, one may use one or more of the following procedures: characterize the eddy current probe, employ artifact standards, and determine probe sensitivity as a function of coil construction.

Probe Characterization—Several electrical parameters can be used to differentiate eddy current probes with poor sensitivity from those with good sensitivity, such as probe inductance in air, impedance in air, and impedance on aluminum and titanium [27]. Minimum values of the impedance change between air and aluminum and between titanium and aluminum can be set as the performance criteria for a particular type of probe. Thus, the impedance change ΔZ_{Ti-Al} is calculated and compared with a minimum value established for that probe or class of probes. An artifact standard consisting of a pair of metal blocks with tapered holes, one block made of Al and the other Ti (Fig. 11), is used in these tests, and it has several advantages over the more conventional slotted block. The advantages of the conductivity blocks are that they require only a routine impedance measurement, and there are no slot sizes to measure. The data may be taken with any standard impedance measuring instrument with a traceable calibration. However, the conductivity blocks (Fig. 11) are difficult to produce because, for example, the required flatness of the surface is difficult to achieve. Moreover, creation of the tapered holes so that a good probe-to-metal contact can be maintained for a range of probe sizes is particularly difficult to realize in aluminum alloy 7075-T6 with minimum damage to the surrounding material.

This test is used for screening, i.e., separating poorly performing probes from those with acceptable sensitivity and which may be useful for performing conductivity tests. However, one cannot predict the exact signal strength that will be obtained when using that probe on actual defects, and the probe must be calibrated for such specific applications.

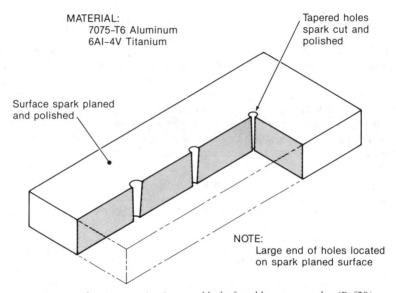

FIG. 11—*Aluminum or titanium test blocks for eddy current probes (Ref 28).*

Artifact Standards—Calibration procedures for eddy current inspection often involve the use of artifact standards containing manufactured flaws, which are assumed to be a good approximation of the type of flaws being sought during inspection. These manufactured flaws are most often produced by electrical discharge machining (EDM), milling, or the controlled growth of fatigue cracks. However, instruments that are sensitive to eddy current signal phase as well as amplitude can show considerable differences between a relatively wide EDM notch or milled slot and a real fatigue crack [29]. The use of controlled growth fatigue cracks also can cause problems when forces at the crack's tip drive the crack faces together, thereby making electrical contact [30]. Moreover, estimates of the crack depth from eddy current measurements based on a calibration using such artificial flaws are difficult to obtain.

A new method for producing artificial flaws that appears to solve some of the difficulties inherent in these conventional techniques was developed [31]. A tightly closed notch is formed by mechanically indenting a specimen, then compressively deforming it until the notch appears to be visibly closed. A reference material containing this type of flaw is available from NIST [32], which provides a reproducible flaw of known size and geometry that closely resembles an actual fatigue crack. In their present form, the notches are fabricated in a 7075-T6 aluminum alloy block (Fig. 12). The size and shape of the notch is controllable and accurately known from the geometry of the indenting tool. The angle between the lift-off impedance vector and the flaw impedance vector is greater for compressed notches than for EDM notches. However, the materials in which the former can be produced are restricted to relatively soft materials, and thus compressed notches in titanium, for example, cannot be made.

Eddy Current Probe Sensitivity as a Function of Coil Construction Parameters—Discussions with eddy current inspectors have indicated that nominally identical probes can have performance (sensitivity) variations of 30% or more. Although a performance specification may be effective in finding probes with poor sensitivity, it seems preferable to develop a tool for predicting performance for a given coil construction and, thus, eliminate coils having low sensitivity. While air core probes can be manufactured with predictable magnetic fields, this is not yet possible for ferrite core probe coils. An empirical approach to solve this problem was

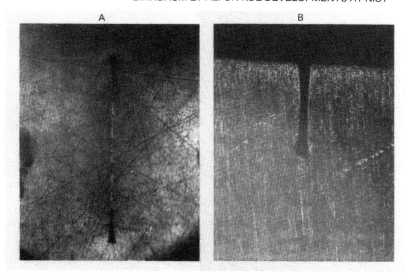

FIG. 12—*Micrographs of compressed notches, standard flaws for eddy current probe characterization: (A) 7075-T6 aluminum alloy, top view; (B) 6061 aluminum alloy, cross-sectional view (Ref 31).*

undertaken [*33*]. The application of computer modeling to the interaction of ferrite core coils with realistic flaw geometries is in its early stage.

A statistical experimental design technique was used to quantify the relationship of eddy current coil construction to the performance and operating characteristics of coils used in NDE inspections [*33*]. The coil chosen for this study was the single-coil, ferrite core type. The six main factors studied were ferrite core diameter, ferrite permeability, coil aspect ratio, number of turns, distance of the windings from the inspection end of the ferrite, and wire gauge. The large number of factors in the design made varying one factor at a time impractical. However, statistical experimental design techniques offered a systematic approach for choosing combinations of factors. It was found that number of turns (NT), winding distance from end (WD), and aspect ratio (AR), as well as two interaction terms (NT)×(WD) and (NT)×(AR) were the dominant factors affecting coil sensitivity, ΔZ. A simple five-term model based on these factors was used to predict ΔZ with reasonable accuracy, as verified by measurements on an EDM notch in an aluminum block.

Uniform Field Eddy Current Probe—Uniform field eddy current (UFEC) probes [*34*] operate by interrogating flaws with a spatially uniform electromagnetic field. A field map of a UFEC probe fabricated at NIST is shown in Fig. 13. Its use in quantitative NDE, as indicated earlier, is particularly attractive because the theoretical models of the field-flaw interactions are greatly simplified and lead to a simpler method for determining flaw sizes from measurements. Auld and coworkers [*35,36*] developed a theory for the interaction of a uniform interrogating field with three-dimensional surface flaws in the limit of small skin depth. An extensive series of measurements was undertaken to evaluate in detail the use of UFEC probes [*37*] for quantitative NDE. The study included a series of semielliptical EDM slots and fatigue cracks in Ti-6Al-4V, from which the following conclusions emerged: calibration of a UFEC probe can be carried out with several calibration artifacts of different geometries; measurements on all EDM slots and fatigue cracks less than 1.5 mm long were in excellent agreement with the prediction of Auld's theory, and a simple method for inverting UFEC measurements to obtain flaw depth when the length is known was demonstrated.

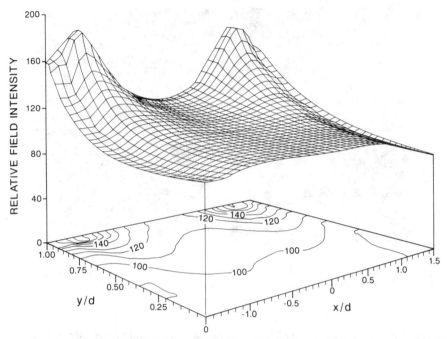

FIG. 13—*Two-dimensional map of the relative magnetic-field intensity between the poles of the NIST uniform field eddy current probe. One half of the area between the two poles is mapped from the center of the probe in the foreground to the tip of the pole in the background (Ref 37).*

The current status of eddy current calibrations and standards may be summarized as follows:

1. The measurement of $\Delta Z_{\text{Ti-Al}}$ reliably screens probes for low and unacceptable sensitivity.
2. The compressed notch is a reliable and reproducible artifact standard for Al but cannot be fabricated for high-strength materials such as Inconel and titanium.
3. EDM notches in certain materials, such as 7075-T6 Al, produce large differences in eddy current response for nominally identical notches.
4. Probe sensitivity models can be useful guides for probe manufacturing.
5. Probe magnetic field measurements are in agreement with theory.
6. Uniform field eddy current probes can quantitatively size EDM slots and fatigue cracks and give results in excellent agreement with theory in the limit of small skin depth ($a/\delta \gg 1$, where a is flaw depth and δ is skin depth).

Area Averaging Techniques For Surface Roughness

Optical Scattering—Concern with surface roughness arises in two general areas: appearance of products such as manufactured household appliances and automobiles, and the functioning of a variety of components. The latter category includes the imaging quality of optical elements, the hydrodynamic and aerodynamic drag of surfaces, and the friction of moving parts. Thus there is a need to measure and monitor surface roughness. The scattering of optical radiation, such as that provided by a laser beam, provides an approach for dealing with this problem.

For very smooth surfaces where the heights of the surface asperities are much smaller than the illuminating wavelength, the surface behaves essentially as a mirror and the vast majority of the scattered light is found in the specular direction. As the surface roughness increases, the intensity of the specular beam decreases markedly [38], and the angular width of the diffusely scattered radiation tends to increase. The angular distribution of the diffusely scattered radiation is closely related to the Fourier decomposition of surface spatial wavelengths [39]. Short surface wavelengths tend to scatter the radiation into wide angles from the specular direction, whereas long wavelengths scatter the radiation close to the specular direction.

In principle, the measurement of various properties of the scattered light can be used, with the appropriate theory, to assess average properties of the surface roughness over the area illuminated by the radiation; such measurements are classified as area techniques for assessing surface roughness. Although area techniques have the advantage that a single or small number of measurements can yield knowledge about the surface roughness without the need for detailed profiling of the surface, the area-averaged parameters cannot be directly related to conventional surface parameters obtained by profiling techniques [40]. Scattering from smooth surfaces has been studied both experimentally and theoretically and is well understood [41]. For machined surfaces where the asperity heights are a sizable fraction of the illuminating wavelength, the scattered light distribution is a complicated function of surface topography, although certain surface texture information may be derived from the scattered light.

The light-scattering instrument developed at NIST, called DALLAS [42,43] uses a He-Ne laser with a wavelength of 0.6328 μm to illuminate the surface. The pattern of scattered radiation is measured by an array of 87 detectors located in a semicircular arc and rotatable over most of the hemisphere above the surface. Excellent agreement between theory and experiment was obtained for all members of a set of nine hand-lapped stainless steel specimens with rms surface roughnesses, R_q, determined from stylus measurements, ranging from 0.08 to 0.48 μm. Note that rms surface roughness is designated by R_q when determined from stylus measurements and by σ when determined from optical scattering. Figure 14 shows the results for a rough (Specimen 6) surface and a smooth (Specimen 14) surface. The scattering pattern for the smooth specimen is characterized by a strong narrow specular peak containing about 59% of the scattered light flux, and a diffuse scattering distribution. Such a surface looks like a clouded mirror with marks and imperfections. For the rough surface, the specular peak is gone and the distribution consists entirely of diffusely scattered light. For both these cases and the other seven of the set, the calculated distributions, based on optical scattering theory whose input is the measured surface profiles, agree reasonably well with the measured data. The differences between theory and experiment are likely due to the lateral resolution limit of the stylus tip width of approximately 0.5 μm. Structures on the surface more closely spaced than this are not well resolved by the stylus tip and tend to scatter the light into wide angles.

These results show that scattering theory can be used to quantitatively describe the scattered light distribution from a knowledge of surface topography. However, for this theory to be useful for NDE, it is necessary to solve the inverse problem: how to use the scattered light distribution to determine surface roughness parameters. This requires that a statistical model of the surface with appropriate geometrical parameters such as the rms surface roughness, σ, and the surface correlation length, T, be folded into a theoretical description of the light scattering. Unfortunately, it was found that although good fits could be obtained to a theoretical expression in terms of the two parameters σ and T, the determination proved to be ambiguous, i.e., several sets of values of σ and T gave equally good fits to the data. Nevertheless, the relative intensity of the specular peak and the angular width of the scattered light distribution appeared to be useful indicators of the surface roughness. In fact, different parameters related to the surface roughness can be determined in different ranges of σ/λ, the ratio of rms surface roughness to optical wavelength, as illustrated in Fig. 15.

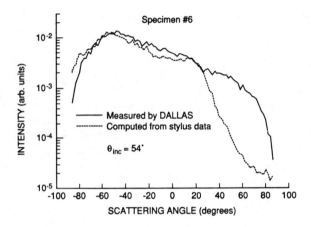

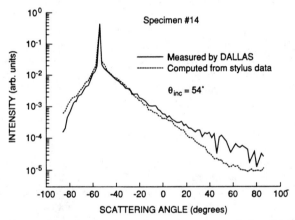

FIG. 14—*Comparison of theoretical and experimental angular scattering distributions measured by DALLAS [42,43] for two stainless steel hand-lapped surfaces (Ref 14).*

For smooth surfaces such that $\sigma/\lambda < 0.01$, the diffraction of the scattered light is a first-order phenomenon, that is, an exponential in an equation of the scattered light can be approximated by an expansion to the first order in σ/λ. As a result, there is a direct mapping between the power spectrum of surface spatial wavelengths and the angular distribution of scattered light [44]. When the above relationship breaks down, other surface roughness quantities can still be determined from the optical scattering. At values of σ/λ less than about 0.14, the height autocorrelation function (ACF) could be deduced from a Fourier transform of the angular distribution of the scattered light, in good agreement with the ACF obtained from the stylus data [45]. However, this procedure requires a value of the rms roughness, which may be separately determined from the relative intensity of the specular beam.

It is well known that the intensity of the specular peak relative to the total amount of scattered light is related to the rms surface roughness [46]. Assuming a Gaussian distribution of surface heights, a relation can be derived allowing one to predict σ, not necessarily the same

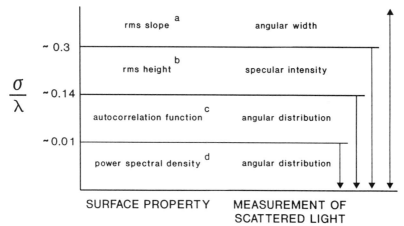

SURFACE PROPERTY   MEASUREMENT OF
SCATTERED LIGHT

FIG. 15—*Diagram illustrating the measureable quantity in the right-hand column as a function of σ/λ, where σ is rms surface height and λ is the wavelength of light. For each band, the calculated surface property on the left, as well as all others above, can be determined. The measurement methods are described in a (Ref 39), b (Ref 45), c (Refs 40,46), and d (Refs 44,47).*

as R_q, from a knowledge of the intensity of the specular peak, as shown in Fig. 16. An analysis of this technique indicates that it serves as a direct indicator of optical roughness for values of σ/λ up to about 1/3, assuming that the angle of incidence is not close to the grazing angle.

In addition, several investigators have shown that the width of the angular scattering distribution is related to the rms slope, Δ_q, of the scattering surface (see, for example, Ref 44). Using the stylus measured values of Δ_q, a linear relationship was found that relates Δ_q to the mean square width of the light-scattering distribution. This linear relationship was found to hold for a number of rough surfaces in agreement with the theoretical predictions of Ref 44. Furthermore, it was found to be applicable for a variety of experimental conditions [47]. These observations are significant because they mean that scattered light can be used to perform direct, on-line, noncontacting measurements of the rms slope of a manufactured rough surface.

Parallel Plate Capacitance Gauge—A comparison of roughness measurements performed on a large variety of metal surfaces using capacitance roughness gauges and high-quality stylus instruments was made [48]. The probing element of the capacitance gauges consists of a 2.00 by 16.88-mm flexible platen, which, together with the metal surface, forms an electrical capacitor, as illustrated in Fig. 17. As mentioned earlier, area-averaging techniques such as surface capacitance yield parameters which depend on the statistical properties of the surface roughness sampled over the entire area of the probe and require a physical model to relate measurements to surface roughness. Modeling the performance of the probe involved an understanding of how each element of the rough surface affects the capacitance and the way the flexible platen rests on top of the highest peaks and sags between these peaks. The model for the behavior of such a probe uses a digitized stylus-generated profile as input data and yields a roughness parameter, R_c. This parameter is the equivalent uniform spacing between a supposedly flat platen and the mean plane of the rough surface. The model was validated by comparing the computed value of R_c with that measured using the capacitance gauge for each of 41 different surfaces. The data of Fig. 18 indicate that the capacitance-based surface parameter R_c correlates with the calculated R_c. This parameter is a better estimator of the mean surface height than the rms surface roughness, R_q.

A subject for standardization is the probe geometry and material constants which can yield different results for R_c. It is possible that the probe compliance can change with age or use.

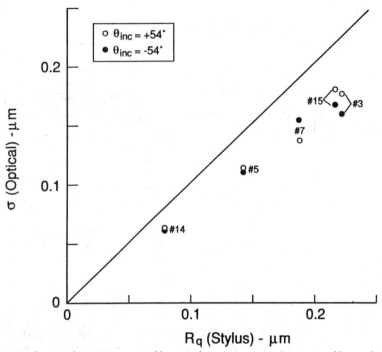

FIG. 16—*Surface roughness (σ) measured by optical scattering versus Rq measured by a stylus for hand-lapped stainless steel surfaces (Ref 41).*

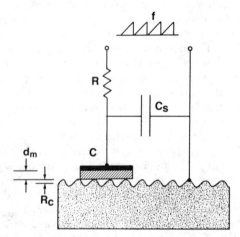

FIG. 17—*Schematic diagram of a capacitance instrument probing the topography of a rough surface. Capacitor C consists of the metallization (black) and the dielectric film (shaded), thickness d_m, resting on the rough metal surface (speckled) beneath it (Ref 48).*

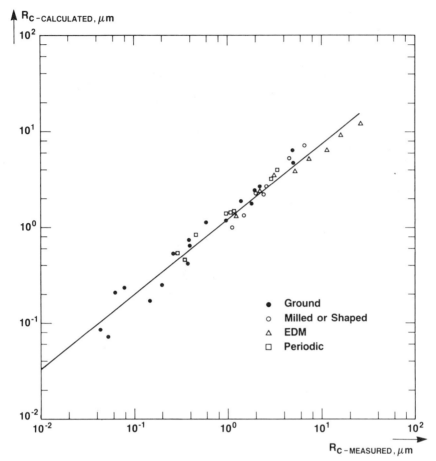

FIG. 18—*Calculated* R_c *versus measured* R_c *for 41 rough surfaces. The least-squares straight-line fit for the log-log data is also shown (Ref 48).*

Therefore, this quantity should be defined and controlled to ensure repeatable operation of the capacitance instrument.

Magnetic Methods

Magnetic Particle Inspection

Although the nondestructive evaluation of magnetic steels by magnetic particle inspection (MPI) has long been considered a mature technology, a number of questions remain on how to obtain reproducible, quantitative results. It is possible to make the method too sensitive and an obscuring background may form, or not sensitive enough and important defects may be missed. The primary factors that must be controlled to obtain reproducible and predictable results are: magnetization level; concentration, shapes, and magnetic properties of the particles; method of particle application; and method of illumination and interpretation of the indications. A number of these factors were recently addressed [49,50]. Here we emphasize the problem of obtaining the correct magnetization, which is not one of making the technique sufficiently sensitive, but one of adjusting the sensitivity into the correct range.

The method of magnetizing the part to be tested and the level of magnetization to be used are always major problems when developing a magnetic particle inspection procedure for a given part. In many cases the leakage field resulting from the magnetization can be well approximated by a single dipole whose field is given by a very simple equation. However, more important than the leakage field values themselves are the forces they produce on magnetic particles. A heuristic approach is used to normalize the force normal to the surface to unity under conditions when it is known that magnetic particle indications will form (the Betz ring [49] is used for this test). An example of the leakage field and normalized forces appropriate for a 1 mm-diameter cylindrical defect centered 1 mm below the surface for an applied field of 43 Oe is shown in Figs. 19a and 19b. According to the normalizing criterion used in this case, this defect should be detectable at fields down to about 5 Oe. Indeed, very small defects which are close to the surface are readily detectable at low fields, providing only that the width of the leakage field is large compared with the particle diameters and that the coercive force of the steel is small. However useful, this detection criterion is in need of many refinements and may be considered no more than an order of magnitude estimate.

The pertinent component of the field for the formation of indications is parallel to the surface of the part and perpendicular to the axis of the defect, i.e., the tangential field. This field can be measured by a small Hall probe at any location on a part where defects must be detected. Figure 20 shows the parameters important in placement of the probe [51]. To measure the field at the surface, the lift-off of the probe and its dimensions should be as small as possible. To ensure that the tangential field is being measured, the plane of the probe must be perpendicular to the surface. (The tangential field is indicated by H_0 in Fig. 19a.)

A magnetic particle testing ring (the so-called Ketos ring) is used for qualifying particles or verifying overall MPI system performance. Up to now, nonreproducible results have been obtained from these rings, mostly due to variability in material. However, 30 new rings are being fabricated at NIST [52] using a new material (vacuum remelted grade 52100 steel), and the perpendicular component of the leakage field (actually the field gradient) is being measured

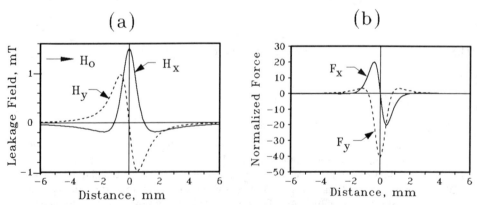

FIG. 19—(a) *Leakage field calculated for a 1-mm-diameter cylindrical defect centered 1 mm below the surface with an applied tangential field H_0 of 4.3 mT (Ref 50). (b) Normalized force on a magnetic particle for the leakage field of Fig. 19a (Ref 50).*

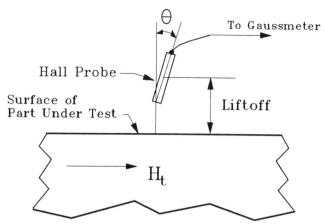

FIG. 20—*Placement of a Hall probe used for measurement of the tangential component of the applied field (Ref 51).*

with a Hall probe to test the performance of each of these rings. Figure 21 shows the leakage fields measured on a ring with twelve holes at various depths. However, seven holes are planned for the new rings. Leakage field measurement is an important test since it serves to prove the performance of the ring without the use of MPI and the uncertainties introduced therein.

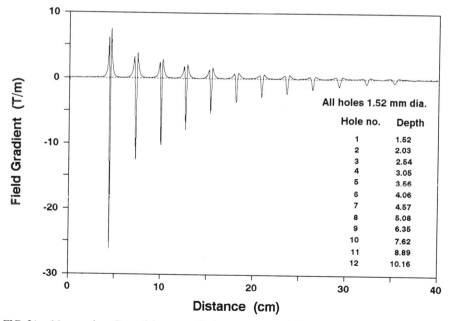

FIG. 21—*Measured gradient of the perpendicular component of the leakage field from a test ring. The ring was fabricated from 52100 steel. The central conductor current was 1400 A (L. J. Swartzendruber).*

Barkhausen Noise

The dynamics of domain wall motion for slowly varying magnetic fields in many ferromagnetic materials is dominated by the Barkhausen effect. This consists of a large number of small irreversible jumps in magnetization due to the pinning of domain wall structures by defects in the material. Measurements of Barkhausen noise are useful for obtaining information about microstructure and residual stress in ferromagnetic materials. The equipment basically consists of a magnet to sweep the sample through a hysteresis loop, a pickup coil which is placed on the surface, a low-noise preamplifier, and a digitizing oscilloscope or equivalent equipment. As the applied field is swept, jump voltage spikes are obtained as a function of the applied field, as shown in Fig. 22. These data can be analyzed in various ways, but the introduction of a jumpsum spectrum as illustrated in Fig. 23 is particularly useful. The jumpsum is simply the sum of the voltage spikes that have occurred up to a given value of magnetic field. The sharp rise in the jumpsum occurs in the steeply varying portion of the magnetization curve. The interesting result in Fig. 23, which records the results for samples with different heat treatments, is that Samples 23 and 28 with vastly different heat treatments registered the same Rockwell hardness (HRB 94), whereas the samples have different tensile strengths, 95 000 and 99 000 psi. However, the jumpsum test easily shows a difference between the two samples. In order to make such tests comparable with different Barkhausen equipments, it is necessary to calibrate them with a suitable standard reference material (SRM) and normalize the results accordingly [54]. The calibration or jumpsum normalization spectrum that is provided by an SRM is shown in Fig. 23. However, the individual jumpsums have not been normalized in this figure.

Miscellaneous Methods

Infrared Thermography

Infrared thermography is the process of displaying variations of apparent temperature over the surface of an object or a scene by measuring variations in infrared radiance. This definition of infrared thermography was crafted by ASTM Subcommittee E07.10 on Emerging NDT Methods, and it appears in a glossary of terms relating to NDT by infrared thermography[2].

NIST personnel played a major role in the development of this glossary, which has been a rather formidable task that is still going on. The difficulties arose from the fact that infrared thermography is a measurement method with many useful applications that predate its more recent emergence as an important NDT method [55]. Aside from a terminology which was not well suited for NDT, the historical uses of infrared thermography led to the evolution of performance criteria for thermographic imaging systems that likewise are not entirely adequate for NDT purposes. In other words, the criteria used by the manufacturers and vendors of thermographic imaging systems to describe the capabilities of their products have not been those which a prospective purchaser would need to know in order to select the best system for a specific NDT application.

A project was established to define the important performance criteria for thermographic imaging systems intended for use in NDT applications and to develop standard test methods for quantifying those criteria. Three such criteria were identified [56]. The first is *minimum resolvable temperature difference*, or MRTD, which is a measure of the ability of an infrared imaging system and a human observer together to recognize periodic bar targets on a display. This characteristic relates to the system's effectiveness for discerning details in a scene. The standard test method, which was developed by NIST in order to measure this quantity, was

---

[2] ASTM E 1149: Terminology Relating to NDT by Infrared Thermography.

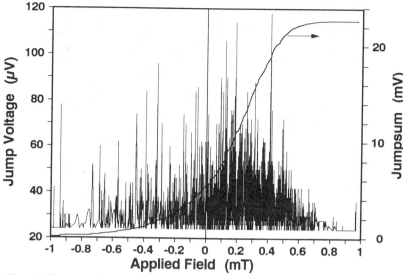

FIG. 22—*Barkhausen noise jumps (left scale) and the sum of those jumps (right scale) as a function of applied field observed in a sample of low carbon steel (Ref 53).*

promulgated as ASTM Standard E 1213[3]. The test method employs a standard four-bar target in conjunction with a differential blackbody that can establish one blackbody isothermal temperature for the set of bars and another blackbody isothermal temperature for the set of conjugate bars that are formed by the regions between the bars (see Fig. 24). The target is imaged

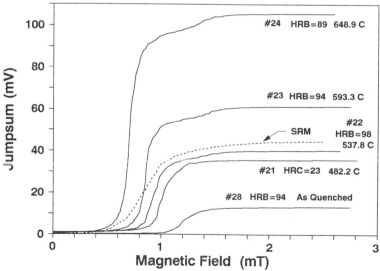

FIG. 23—*Jumpsum spectra of ASTM A710 steel, austenized for 90 min at 900°C, quenched, and then aged for 90 min at the indicated temperatures (Ref 54).*

[3] ASTM E 1213: Test Method for Minimum Resolvable Temperature Difference for Thermal Imaging Systems.

FIG. 24—*Targets used for minimum resolvable temperature difference (MRTD) determinations (ASTM E 1213–87). Targets may be fabricated by cutting slots in metal and coating with black paint.*

onto the video monitor of the thermal imaging system being evaluated. The video image is viewed by the observer as the temperature difference between the bars and their conjugates, initially zero, is increased incrementally. The temperature difference at which the observer can first distinguish the four bars is the MRTD.

The second performance criterion for thermal imaging systems intended for NDT applications is *minimum detectable temperature difference*, or MDTD. This is a measure of the compound ability of the system and the observer to detect a target of unknown location at one temperature against a background at another temperature. This quantity relates to the detection of small material defects such as cracks, voids, or inclusions. E 1311 is the standard test method developed by NIST and adopted by ASTM for measuring this property[4]. The image employed in this test method consists of a small circular target at one blackbody isothermal temperature against a large background at a second blackbody isothermal temperature. The location of the target relative to the background is not disclosed to the observer, who views the image on the video monitor as the temperature difference between the target and the background, initially zero, is increased incrementally until the observer, in a limited duration, can just distinguish the target. The temperature difference at this point is the MDTD.

Noise equivalent temperature difference, or NETD, is the third performance criterion. This is the target-to-background temperature difference between a blackbody target and its blackbody background at which the signal-to-noise ratio of a thermal imaging system is unity. This provides an objective measure of the temperature sensitivity of the system. A proposed test method for measuring this quantity was developed by NIST and is presently making its way through the ASTM balloting process. In this test method, a specified temperature difference is established between a blackbody target and its blackbody background, and the NETD is calculated from measurements of the peak-to-peak signal voltage from the target and the rms noise voltage from the background.

Leak Testing

A program to develop leak rate standards (i.e., low flow-rate standards) at NIST evolved from an existing core of competence in high-vacuum measurements [57]. The primary leak

[4] ASTM E 1311: Test Method for Minimum Detectable Temperature Difference for Thermal Imaging Systems.

standard developed at NIST [58] (see Fig. 25) consists of three major components: the vacuum chamber, the leak manifold, and the flowmeter. The critical element of the system is the constant-pressure, variable-volume, low-range flowmeter [59], which was developed specifically for use with vacuum and leak rate standards.

In operation, the leak artifact to be calibrated is placed on the manifold (Fig. 25) and the entire system is evacuated. The flowmeter is isolated from the rest of the system, and gas from the unknown leak is allowed to flow into the vacuum chamber, through an orifice, and is then evacuated by the pump. After the gas flow and the upper chamber pressure reach steady values, the upper chamber pressure indication is recorded. The leak manifold is then shut off from the vacuum chamber, the flowmeter is pressurized with the gas species of interest, and the flow is adjusted so that when everything is stabilized the upper chamber pressure indication is the same as when the flow from the leak was passing through the system. The flow rate is now measured by shutting off the variable volume from the reference volume and activating a servomechanism (driven by a capacitance diaphragm gauge) that drives the piston into the variable volume at a rate that will keep the pressure in that volume constant as gas flows into the vacuum chamber. The flow rate into the main chamber can be calculated from the rate at which the piston is advanced, the cross-sectional area of the piston, and the pressure and temperature of the gas in the variable volume. Within limits determined by the instabilities of the upper chamber pressure gauge, this flow rate is equal to the flow rate of the unknown leak.

This system allows the measurement of leak flow rates over the range from 10^{-8} to 10^{-14} mol/s (2×10^{-4} to 2×10^{-10} atm-cm^3/s at 0°C) with an uncertainty between 2 and 10%. Calibration services for reference leaks, which are used as transfer standards, are offered over this range and include a determination of temperature dependence from 0 to 50°C.

These calibrations are currently limited to helium permeation leaks. The calibration range and the demand for this service have both increased significantly since the service was first introduced in 1987, but there is also considerable demand to extend leak rate calibrations to gases other than helium, particularly to freons and sulfur hexafluoride. While the measurement standard is capable of operating with other gases, data on the stability of transfer leaks for other gases are presently incomplete and inconsistent.

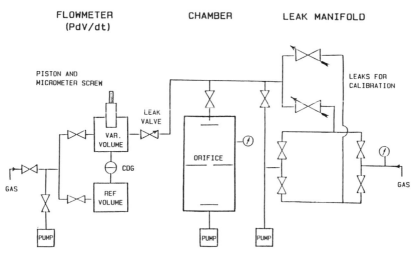

FIG. 25—*Schematic arrangement of primary leak testing standard at NIST showing the flow meter, vacuum chamber, and leak manifold on which leak artifacts to be calibrated are mounted (Ref 58).*

Research and development activities are underway to investigate the characteristics of some other types of transfer leaks, such as sintered plugs [58,60], and experiments will soon begin to examine the stability of transfer leaks for other gases.

Visual Acuity

"Visual acuity is vital to the first step in the three-step interpretation process: (1) detection; (2) interpretation; and (3) evaluation. Individual visual acuity can and does vary from day to day depending on physiological and psychological factors." [61]

The development of a new standard test method to assess the visual acuity of inspectors and others who read and interpret radiographs is being pursued by ASTM Subcommittee E07.02 on Reference Radiographs. The basis of the proposed standard is a set of reference radiographs of artifacts containing tight discontinuities (artificial cracks) of different widths in different locations, orientations, contrasts, and levels of blur (Fig. 26).

This concept for a visual acuity test method was formulated a decade ago [62]. The meticulous fabrication of the artifacts was subsequently carried out over a period of years as part of

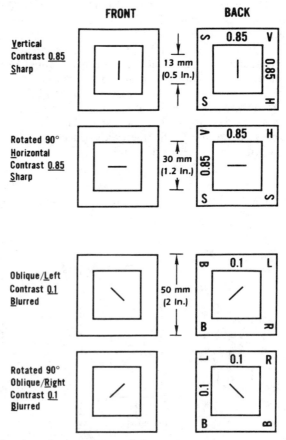

FIG. 26—*Examples of visual acuity test images showing line orientations and dimensions (Ref 62).*

NIST's project to develop new and improved standards for nondestructive testing and inspection for military applications[5].

The Subcommittee arranged for several sets of transparencies to be made from the reference radiographs, and these were circulated to participating organizations for a round-robin evaluation, which revealed that the images of many of the discontinuities are too readily seen to provide adequate discrimination. The original set of 72 artifacts has, therefore, been pared down and consideration is being given to making some of the remaining images less detectable.

In spite of this obstacle, interest in this subject continues to run high and it is now anticipated that a visual acuity test method of this kind would also prove useful to practitioners of NDT disciplines other than radiography, such as visual inspection and liquid penetrant testing, in which visual acuity is clearly a critical factor.

X-Ray Radiology

NIST's role in the development, maintenance, and dissemination of measurement standards for X-ray radiology has been fulfilled by the active participation of NIST personnel in the activities of ASTM Subcommittee E07.01 on the Radiology (X and Gamma) Method. This participation involved the leadership of task groups, drafting of document standards, formulation of test protocols, and the conduct of round-robin test programs, as well as the myriad of other meticulous tasks that must be pursued with diligence in order to harmonize honest differences of opinion among the members of a large committee regarding preferred technical approaches.

A noteworthy example of the ASTM standards that emerged from these collaborative efforts is E 746, Method for Determining Relative Image Quality Response of Industrial Radiographic Film. This test method employs a special image quality indicator (IQI) designed at NIST (Fig. 27). The first embodiments of this large and complex IQI were made available by NIST through its Standard Reference Materials Program; detailed dimensional information provided in the standard now enables acceptable replicas of the artifact to be fabricated commercially.

The test method generates a "classification index" for each film type tested. This quantity is defined as the equivalent penetrameter sensitivity of the film at which exactly half of the holes in the IQI are visible to readers. Equivalent penetrameter sensitivity (EPS), in turn, is defined as

$$\text{EPS},\% = (100/X)(hT/2)^{1/2}$$

where h = hole diameter, mm, T = step thickness of IQI, mm, and X = thickness of test object, mm.

The development of this standard was a major step forward in the standardization of radiographic NDE practices, and this document is now under consideration by the International Organization for Standardization[6].

In recent years the main thrusts of NIST's standardization efforts in X-ray radiology have shifted from conventional film radiography to real-time radioscopy and laminography, while retaining the close association with ASTM [63].

---

[5] This project, which has received long-term support from the Army's Materials Technology Laboratory in Watertown, MA, has produced a number of significant new military and ASTM standards.
[6] ISO/TC 135/SC 5 N64: NDT. Radiographic Inspection. Determining Relative Image Quality Response of Industrial Radiographic Film.

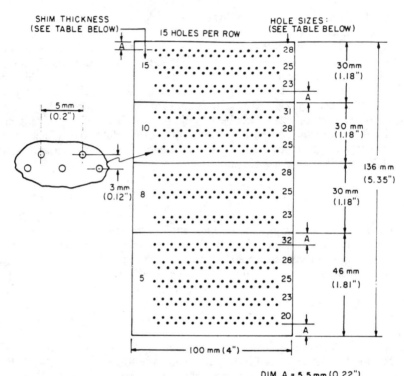

DIM. A = 5.5 mm (0.22")

Step Identification	Shim Thickness, mm (in.)	Hole Identification	Hole Size, mm (in.)
15	0.38 ± 0.012 (0.015 ± 0.0005)	32	0.81 ± 0.025 (0.032 ± 0.001)
10	0.25 ± 0.012 (0.010 ± 0.0005)	31	0.79 ± 0.025 (0.031 ± 0.001)
8	0.20 ± 0.012 (0.008 ± 0.0005)	28	0.71 ± 0.025 (0.028 ± 0.001)
5	0.13 ± 0.012 (0.005 ± 0.0005)	25	0.64 ± 0.025 (0.025 ± 0.001)
		23	0.58 ± 0.025 (0.023 ± 0.001)
		20	0.50 ± 0.025 (0.020 ± 0.001)

Hole Spacing (horizontal): 5 ± 0.1 mm (0.2 ± 0.004 in.) Nonaccumulative
Row Spacing: 3 ± 0.1 mm (0.12 ± 0.004 in.)
Spacing between hole sets: 5 ± 0.1 mm (0.2 ± 0.004 in.)
All other dimensions shall be in accordance with standard engineering practice.

FIG. 27—*Special image quality indicator for X-ray radiography (ASTM E 746–87).*

Radioscopy uses X-rays, like film radiography, but replaces the film with an electronic imaging system. The NIST effort is leading to new devices and test procedures for measuring the special imaging capabilities of real-time radioscopy, such as the examination of objects while they are in motion. One of these new devices is a large steel plate with various IQIs on its surface. The IQIs include conventional radiographic penetrameters as well as line-pair gauges and other research-oriented items. While the penetrameters should be suitable for measuring most radioscopic image parameters in a static mode, they are not expected to be useful for quantifying image quality or system performance during translation or rotation, two of the special capabilities of a real-time system. The line-pair gauges and a hollow sphere design [64], Fig. 28, are being evaluated for use with these capabilities. Line-pair gauge resolution may be expected to degrade when translated normal to the X-ray beam, but the hollow sphere design produces the same image regardless of its orientation and can perceive translation in any direc-

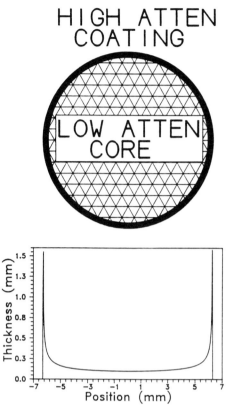

FIG. 28—*A rotationally invariant image quality indicator for X-ray radiology, fabricated by coating a sphere of low-attenuation material with a thin layer of high-attenuation material. The projected thickness of the coating is plotted in the lower portion of the figure (Ref 64).*

tion by blurring of its edges. While it can be used as a penetrameter, the hollow sphere has the advantage of a more conspicuous image (because the edges are very dark) than that of a conventional flat penetrameter. It has been suggested that the best application of the hollow sphere may be for the measurement of unsharpness in the image.

A military standard for radioscopic inspection is being developed by NIST in cooperation with the General Electric Company and the U.S. Army Materials Technology Laboratory. This document is intended to be fully compatible with ASTM E 1255[7] and to replace or strengthen its procedure section. The new document will prescribe radioscopic inspection requirements for all materials and, taken together with ASTM E 1255, will set forth the criteria for qualifying radioscopic systems and monitoring their day-to-day operation. The goal of this work is to facilitate the replacement of radiographic inspection by radioscopic inspection when equal or better inspection can thereby be obtained.

Laminography is an imaging technique that can capture two-dimensional slices through a three-dimensional object. Its advantage over conventional radiography is its ability to image individual planes, say, in a part situated beneath another part. In a project supported by the U.S. Army's Harry Diamond Laboratories, NIST scientists are assessing the advantages and

[7] ASTM E 1255: Standard Practice for Radioscopic Real Time Examination.

disadvantages of a commercial laminography system intended for the inspection of solder connections on printed wiring boards. Here, too, special test objects are being developed in order to permit quantitative measurements, such as a calibration board constructed with holes of various diameters on the X- and Y-coordinates [65]. Preliminary tests have confirmed the ability of the calibration board to yield meaningful data on the inspection system's magnification and positioning accuracy. A batch of the calibration boards are being produced and certified. They will be available from NIST's Standard Reference Materials Program as SRM 1842. A similar concept is being developed for X-ray systems with z-dimension (laminographic) capabilities. A wedge-shaped device has been constructed from alternate layers of materials that absorb and transmit X-rays. The device has demonstrated a resolution of 0.2 mm with one laminographic system, and an improved version is being constructed. This device is expected to become a standard reference material in the next year. A procedure to evaluate the system's reproducibility in terms of line-pair resolution is under development.

Concluding Remarks

Although much progress has been made in developing a very wide variety of standards for determining the reliability and accuracy of NDE equipment and measurements, and, indeed, certain areas appear to have reached some level of maturity, it is also clear that much remains to be done. The standards that are discussed here were developed primarily with the needs of postmanufacturing and in-service inspection in mind. However, there is a growing interest and need for developing NDE instruments for sensing various aspects of materials during processing, and, of the manufacturing process itself. Such instruments may have to operate in the harsh environment of a manufacturing facility and, moreover, operate continuously or nearly so. Consequently, it appears that they could require even more frequent and stringent calibration than is associated with NDE for postmanufacturing and in-service inspection. To date, little attention appears to have been devoted to these problems. We expect this situation to change in the near future.

Acknowledgments

The authors thank T. E. Capobianco, L. J. Swartzendruber, and T. V. Vorburger for their assistance in preparing, respectively, the sections on eddy currents, magnetic methods, and area averaging techniques for surface roughness. The authors also thank F. Mopsik for a critical reading of the manuscript.

References

[*1a*] Cloeren, L. C. S. and Eitzen, D. G., "Ultrasonic Standards Group Publications," NIST-LP-102, National Institute of Standards and Technology, Gaithersburg, MD, 1991.

[*1*] Breckenridge, F. R., "Acoustic Emission Transducer Calibration by Means of the Seismic Surface Pulse," *Journal of Acoustic Emission*, Vol. 1, No. 2, 1982, pp. 87–94.

[*2*] Proctor, T. M., Jr., "More Recent Improvements on the NBS Conical Transducer," *Journal of Acoustic Emission*, Vol. 5, No. 4, 1986, pp. 134–142.

[*3*] Proctor, T. M., Jr., "A High Fidelity Piezoelectric Tangential Displacement Transducer for Acoustic Emission," *Journal of Acoustic Emission*, Vol. 7, No. 1, 1988, pp. 41–47.

[*4*] Greenspan, M., "The NBS Conical Transducer: Analysis," *Journal of the Acoustic Society of America*, Vol. 81, No. 1, 1987, pp. 173–183.

[*5*] Breckenridge, F. R., Proctor, T. M., Jr., Hsu, N. N., and Eitzen, D. G., "Some Notions Concerning the Behavior of Transducers," *Progress in Acoustic Emission III, Proceedings of the Eighth International Acoustic Emission Symposium*, K. Yamaguchi, K. Aoki and T. Kishi, Eds., The Japanese Society of NDI, Tokyo, Japan, 1986, pp. 675–684.

[6] Breckenridge, F. R. and Greenspan, M., "Surface-Wave Displacement: Absolute Measurements Using a Capacitive Transducer," *Journal of the Acoustic Society of America*, Vol. 69, No. 4, 1981, pp. 1177–1185.

[7] Breckenridge, F. R., Proctor, T. M., Jr., Hsu, N. N., Fick, S. E., and Eitzen, D. G., "Transient Sources for Acoustic Emission Work," *Progress in Acoustic Emission V, Proceedings of the Eighth International Acoustic Emission Symposium*, K. Yamagouchi, H. Takahashi, and N. Niitsuma, Eds., The Japanese Society for NDI, Tokyo, Japan, 1990, pp. 20–37.

[8] Eitzen, D. G., Breckenridge, F. R., Clough, R. B., Fuller, E. R., Hsu, N. N., and Simmons, J. A., "Fundamental Developments for Quantitative Acoustic Emission Measurments," NP-2089, Research Project 608–1, Electric Power Research Institute, Palo Alto, CA, October 1981.

[9] Pao, Y. H., Gajewski, R. R., and Ceranoglu, A. N., "Acoustic Emission and Transient Waves in an Elastic Plate," *Journal of the Acoustic Society of America*, Vol. 65, No. 1, 1979, pp. 96–105.

[10] Hsu, N. N., "Dynamic Green's Functions of an Infinite Plate—A Computer Program," NBSIR 85–3234, National Bureau of Standards, Washington, DC, August 1985.

[11] Proctor, T. M., Jr., "Improved Piezoelectric Transducers for Acoustic Emission Signal Reception," *Journal of the Acoustic Society of America*, Vol. 68, Supplement 7, 1980, p. S68.

[12] Proctor, T. M., "A High Fidelity Piezoelectric Tangential Displacement Transducer for Acoustic Emission," *Journal of Acoustic Emission*, Vol. 7, No. 1, 1988, pp. 41–48.

[13] Hsu, N. N., "Acoustic Emission Simulator," U.S. Patent 4018084, May 1976.

[14] Higo, Y. and Inaba, H. "Characterizations of Pencil Lead for AE System Calibration," *Progress in Acoustic Emission, IV*, K. Yamagouchi, I. Kimpara, and Y. Higo, Eds., The Japanese Society of NDI, Tokyo, Japan, 1988, pp. 164–169.

[15] Standard Practice for Fabricating and Checking Aluminum Alloy Ultrasonic Standard Reference Blocks, E 127, *Annual Book of ASTM Standards*, Vol. 3.03, ASTM, Philadelphia, 1990, pp. 17–26.

[16] Greenspan, M., Breckenridge, F. R., and Tschiegg, C. E., "Ultrasonic Transducer Power Output by Modulated Radiation Pressure," *Journal of the Acoustic Society of America*, Vol. 63, No. 4, 1978, pp. 1031–1038.

[17] Blessing, G. V., "An Assessment of Ultrasonic Reference Block Calibration Methodology," NBSIR 83–2710, National Bureau of Standards, Washington, DC, June 1983.

[18] Blessing, G. V., Eitzen, D. G., Henning, J. F., Kodoma, R. H., Clark, A. V., and Schramm, R. E., "Precision Ultrasonic Thickness Measurements on Thin Steel Parts," *Materials Evaluation*, Vol. 49, No. 8, 1991, pp. 982–992.

[19] Blessing, G. V. and Eitzen, D. G., "Ultrasonic Sensor for Measuring Surface Roughness," *Surface Measurment and Characterizations*, Jean M. Bennett, Ed., *Proceedings of the International Society for Optical Engineering*, Vol. 1009, 1988, pp. 282–289.

[20] Papers on calibration and standards in *Eddy-Current Characterization of Materials and Structures*, ASTM STP 722, G. Birnbaum and G. Free, Eds., American Society for Testing and Materials, Philadelphia, PA, 1981.

[21] Moulder, J. C., Gerlitz, J. C., Auld, B. A., Riaziat, M., Jeffries, S., and McFetridge, G., "Calibration Methods for Eddy Current Measurement Systems," *Review of Progress in Quantitative Nondestructive Evaluation*, Vol. 4A, D. O. Thompson and D. E. Chimenti, Eds., Plenum Press, New York, 1984, pp. 411–420.

[22] Dodd, C. V. and Deeds, W. E., "Analytical Solutions to Eddy-current Probe-coil Problems," *Journal of Applied Physics*, Vol. 39, No. 6, 1968, pp. 2829–2838.

[23] Auld, B. A., Ayter, S., Munnemann, F., and Riaziat, M., "Eddy Current Signal Calculations for Surface Breaking Cracks," *Review of Progress in Quantitative Nondestructive Evaluation*, Vol. 3A, D. O. Thompson and D. E. Chimenti, Eds., Plenum Press, New York, 1984, pp. 489–498.

[24] Auld, B. A., Munnemann, F. G., and Riaziat, M., "Quantitative Modeling of Flaw Responses in Eddy-current Testing," *Research Techniques in Nondestructive Testing*, Vol. VII, R. S. Sharpe, Ed., Academic Press, London, 1984, pp. 37–76.

[25] Auld, B. A., Munnemann, F. G., and Burkhardt, G. L., "Experimental Methods for Eddy Current Probe Design and Testing," *Review of Progress in Quantitative Nondestructive Evaluation*, Vol. 3A, D. O. Thompson and D. E. Chimenti, Eds., Plenum Press, New York, 1984, pp. 477–488.

[26] Capobianco, T. E., Fickett, F. R., and Moulder, J. C., "Mapping of Eddy Current Probe Fields," *Review of Progress in Quantitative Nondestructive Evaluation*, Vol. 5A, D. O. Thompson and D. E. Chimenti, Eds., Plenum Press, New York, 1985, pp. 705–711.

[27] Capobianco, T. E., "Field Mapping and Performance Characterization of Commercial Eddy Current Probes," *Review of Progress in Quantitative Nondestructive Evaluation*, Vol. 6A, D. O. Thompson and D. E. Chimenti, Eds., Plenum Press, New York, 1987, pp. 687–694.

[28] Dulcie, L. L. and Capobianco, T. E., "New Standard Test Method for Eddy Current Probes," *Proceedings of the 36th Defense Conference on NDT*, Army Aviation Systems Command, St. Louis, MO, 1988, pp. 154–160.

[29] Rummel, W. D., Christner, B. K., and Long, D. L., "Assessment of Eddy Current Probe Interactions With Defect Geometry and Operating Parameter Variations," *Review of Progress in Quantitative Nondestructive Evaluation*, Vol. 6A, D. O. Thompson and D. E. Chimenti, Eds., Plenum Press, New York, 1987, pp. 705–712.

[30] Moulder, J. C., Shull, P. J., and Capobianco, T. E., "Uniform Field Eddy Current Probe: Experiments and Inversion for Realistic Flaws," *Review of Progress in Quantitative Nondestructive Evaluation*, Vol. 6A, D. O. Thompson and D. E. Chimenti, Eds., Plenum Press, New York, 1987, pp. 601–610.

[31] Capobianco, T. E., Ciciora, S. J., and Moulder, J. C., "Standard Flaws for Eddy Current Probe Characterization," *Review of Progress in Quantitative Nondestructive Evaluation*, Vol. 8A, D. O. Thompson and D. E. Chimenti, Eds., Plenum Press, New York, 1989, pp. 985–989.

[32] NIST Standard Reference Material No. 8458.

[33] Capobianco, T. E., Splitt, J., and Iyer, H., "Eddy Current Probe Sensitivity as a Function of Coil Construction Parameters," *Research in NDE*, Vol. 2, Springer-Verlag, New York, 1990, pp. 169–189.

[34] Moulder, J. C., Shull, P. J., and Capobianco, T. E., "Uniform Field Eddy Current Probe: Experiments and Inversion for Realistic Flaws," *Review of Progress in Quantitative Nondestructive Evaluation*, Vol. 6A, D. O. Thompson and D. E. Chimenti, Eds., Plenum Press, New York, 1987, pp. 601–610.

[35] Auld, B. A., Munnemann, F. G., and Winslow, D. K., "Eddy Current Probe Response to Open and Closed Surface Flaws," *J. Nondestructive Evaluation*, Vol. 2, No. 1, 1981, pp. 1–21.

[36] Auld, B. A., Jeffries, S., Moulder, J. C., and Gerlitz, J. C., "Semi-elliptical Surface Flaw EC Interaction and Inversion: Theory," *Review of Progress in Quantitative Nondestructive Evaluation*, Vol. 5A, D. O. Thompson and D. E. Chimenti, Eds., Plenum Press, New York, 1986, pp. 383–393.

[37] Shull, P. J., Capobianco, T. E., and Moulder, J. C., "Design and Characterization of Uniform Field Eddy Current Probes," *Review of Progress in Quantitative Nondestructive Evaluation*, Vol. 6A, D. O. Thompson and D. E. Chimenti, Eds., Plenum Press, New York, 1987, pp. 695–703.

[38] Beckmann, P. and Spizzichino, A., *The Scattering of Electromagnetic Waves From Rough Surfaces*, Pergamon Press, London, 1963.

[39] Church, E. L., Jenkinson, H. A., and Zavada, J. M., "Relationship Between Surface Scattering and Microtopographic Features," *Optical Engineering*, Vol. 18, No. 2, 1979, pp. 125–136.

[40] Marx, E. and Vorburger, T. V., "Direct and Inverse Problems of Light Scattered by Rough Surfaces," *Applied Optics*, Vol. 29, No. 25, 1990, pp. 3613–3626.

[41] Vorburger, T. V., Marx, E., and Kiely, A., "Optical Scattering for Roughness Inspection," *Progress Report of the Quality in Automotive Projects for FY '89*, NISTIR 4322, T. V. Vorburger and B. Scace, Eds., May 1990, pp. 127–145.

[42] Marx, E. and Vorburger, T. V., "Light Scattered by Random Surfaces and Roughness Determination," in *Scatter from Optical Components, Proceedings of SPIE-The International Society for Optical Engineering*, Bellingham, WA, Vol. 1165, 1989, pp. 72–86.

[43] Vorburger, T. V., Teague, E. C., Scire, F. E., McLay, M. J., and Gilsinn, D. E., "Surface Roughness Studies with DALLAS—Detector Array For Laser Light Angular Scattering," *Journal of Research of the National Bureau of Standards*, Vol. 89, 1984, pp. 3–16.

[44] Rakels, J. H., "Recognized Surface Finish Parameters Obtained from Diffraction Patterns of Rough Surfaces," *Surface Measurement and Characteristics, Proceedings of SPIE-The International Society for Optical Engineering*, Bellingham, WA, Vol. 1009, 1988, pp. 119–125.

[45] Marx, E., Leridon, B., Lettieri, T. R., Song, J. F., and Vorburger, T. V., "Autocorrelation of One-dimensionally Rough Surfaces," to be published.

[46] Porteus, J. O., "Relation Between the Height Distribution of a Rough Surface and the Reflectance at Normal Incidence," *Journal of the Optical Society of America*, Vol. 53, No. 12, 1963, pp. 1394–1402.

[47] Cao, Lin-xiang, Vorburger, T. V., Lieberman, A. G., and Lettieri, T. R., "Light-scattering Measurement of the RMS Slopes of Rough Surfaces," *Applied Optics*, Vol. 30, No. 22, 1991, pp. 3221–3227.

[48] Lieberman, A. G., Vorburger, T. V., Giauque, C. H. W., Risko, D. G., and Rathburn, K. R., "Comparison of Capacitance and Stylus Measurements of Surface Roughness," *Metrology and Properties of Engineering Surfaces 1988*, K. J. Stout and T. V. Vorburger, Eds., Kogan Page, London, 1988, pp. 115–130.

[49] Skeie, K. and Hagemaier, D. J., "Quantifying Magnetic Particle Inspection," *Materials Evaluation*, Vol. 46, No. 6, 1988, pp. 779–785.

[50] Swartzendruber, L. J., "Quantitative Problems in Magnetic Particle Inspection," *Review of Progress in Quantitative Nondestructive Evaluation*, Vol. 8B, D. O. Thompson and D. E. Chimenti, Eds., Plenum Press, New York, 1989, pp. 2133–2140.

[51] Strauss, B. and Swartzendruber, L. J., "Magnetic Particle Inspection Using MIL-STD-1949," *Materials Evaluation*, Vol. 48, No. 3, 1990, pp. 331–334.

[52] Personal communication from Dr. L. J. Swartzendruber.

[53] Swartzendruber, L. J., Bennett, L. H., Ettedqui, H., and Aviram, I., "Barkhausen Jump Correlations in Thin Foils of Fe and Ni," *Journal of Applied Physics*, Vol. 67, No. 9, 1990, pp. 5469–5471.

[54] Personal communication from Dr. L. J. Swartzendruber.

[55] Cohen, J., "Fundamentals and Applications of Infrared Thermography for Nondestructive Testing," *International Advances in Nondestructive Testing*, Vol. 13, Gordon and Breach, New York, 1988, pp. 39–81.

[56] Cohen, J., "Thermal-Imaging System Performance Measures for Nondestructive Testing," *An International Conference on Thermal Infrared Sensing for Diagnostics and Control (Thermosense VI)*, Proceedings of SPIE 446, Bellingham, WA, 1984, pp. 176–180.

[57] Tilford, C. R., personal communication.

[58] Hyland, R. W., Ehrlich, C. D., Tilford, C. R. and Thornberg, S., "Transfer Leak Studies and Comparisons of Primary Leak Standards at the National Bureau of Standards and Sandia National Laboratories," *Journal of Vacuum Science and Technology*, Vol. A4, No. 3, Part 1, 1986, pp. 334–337.

[59] McCulloh, K. E., Tilford, C. R., Ehrlich, C. D. and Long, F. G., "Low-range Flowmeters for Use with Vacuum and Leak Standards," *Journal of Vacuum Science and Technology*, Vol. A5, No. 3, 1987, pp. 376–381.

[60] Ehrlich, C. D., Tison, S. A., Hsiao, H. Y. and Ward, D. B., "A Study of the Linearity of Transfer Leaks and a Helium Leak Detector," *Journal of Vacuum Science and Technology*, Vol. A8, No. 6, 1990, pp. 4086–4091.

[61] "Visual Acuity," Section 8, Part 3 of *Nondestructive Testing Handbook*, 2nd ed., Vol. 3, *Radiography and Radiation Testing*, L. E. Bryant and P. McIntire, Eds., American Society for Nondestructive Testing, Columbus, OH, 1985, pp. 382–383.

[62] Yonemura, G. T., "Visual Acuity Testing of Radiographic Inspectors in Nondestructive Inspection," Technical Note 1143, National Bureau of Standards, Gaithersburg, MD, June 1981.

[63] Siewert, T. A., "Improved Standards for Real-Time Radioscopy," *Nondestructive Evaluation: NDE Planning and Application*, NDE-Vol. 5, R. D. Streit, Ed., American Society of Mechanical Engineers, New York, 1989, pp. 95–97.

[64] Siewert, T. A., Fitting, D. W., and Austin, M. W., "Rotationally Symmetric Image Quality Indicator for Digital X-Ray Imaging Systems," *Materials Evaluation*, Vol. 50, No. 3, 1992, pp. 360–366.

[65] Siewart, T. A., Austin, M. W., Lucey, G. K., and Plott, M. J., "Evaluation and Qualification of Standards for an X-Ray Laminography System," to be published.

Bernard Strauss[1]

NDT Standards from the Perspective of the Department of Defense

REFERENCE: Strauss, B., **"NDT Standards from the Perspective of the Department of Defense,"** *Nondestructive Testing Standards—Present and Future, ASTM STP 1151,* H. Berger and L. Mordfin, Eds., American Society for Testing and Materials, Philadelphia, 1992, pp. 126–135.

ABSTRACT: This presentation illustrates the interaction of the DoD non-Government Society (NGS) bodies in the area of NDT. The adoption process for NGS will be outlined including the criteria for adoption, what adoption means, and the advantages of DoD/NGS interaction.

The tasks of the DoD's Standardization Program Plan for NDT will be described along with DoD's efforts on a Joint Army, Navy, Air Force (JANNAF) NDE Subcommittee and on an international standardization group (America, Britain, Canada, and Australia) called the Quadripartite Working Group on Proofing, Inspection, and Quality Assurance.

KEY WORDS: DoD standardization, NDT, nondestructive testing standardization

It is in the best interests of the Department of Defense (DoD) to interact closely with non-Government standards (NGS) bodies in an attempt to develop usable standardization documents. This paper will study that interaction by indicating how the DoD interacts, how it adopts, what type of documents it adopts, what adoption means, and the advantages or working closely with NGS bodies.

The DoD, by means of a Standardization Program Plan for NDT, outlines its tasks for the next two years. In review of these tasks, it will become apparent that most tasks involve NGS bodies. An ideal approach is for DoD to initiate a project, build up a draft document, and then introduce it into an NGS group. Consequently, many DoD documents have been formatted into ASTM form, and hopefully DoD will adopt the ASTM document (depending on the type of NGS changes) and cancel its military document. This paper will also highlight DoD's efforts in JANNAF (Joint Army, Navy, NASA, and Air Force) efforts in nondestructive evaluation. The JANNAF NDE Subcommittee involves mostly persons other than those active in ASTM or SAE work. Also, this paper will highlight DoD's efforts in international standardization, particularly the Army's work in the ABCA (America, Britain, Canada, and Australia) Quadripartite Working Group on Proofing, Inspection and Quality Assurance.

Criteria for Adoption

Adoption by the DoD indicates acceptance. If a non-Government standard (NGS) is adopted, it is listed in the Department of Defense Issue of Specifications and Standards (DOD-ISS), which is available at all DoD standardization offices, and the document is given more

---

[1] Supervisory materials engineer, U.S. Army Materials Technology Laboratory, SLCMT-MEE, Watertown, MA 02172-0001.

formal consideration in the hierarchy of documents for selection in design. In addition, each ASTM document adopted contains the following caption directly under its title: "This practice (standard or other) has been approved for use by agencies of the Department of Defense and for listing in the DoD Index of Specifications and Standards." Documents are adopted to indicate preference for their use and to increase their visibility. Presently documents are adopted by specific issue, and revisions are not automatically adopted. This allows DoD to not adopt a revision containing undesirable changes.

Non-Government documents are adopted when it is feasible, economical, and practical. Two criteria must be met:

1. Is an NGS available which meets or with minor modification can be made to meet all needs of the DoD with respect to technical requirements and policies?
2. Will an NGS be available in time to meet DoD needs?

Documents proposed for adoption by the DoD must be readily available to the DoD and its contractors. The basic requirement is that sufficient copies of documents be available, either purchased or printed with permission, to meet DoD needs, and that documents be available to contractors from the non-Government standards body (NGSB). Many NGSBs have stated availability conditions necessary to the process of having their standards adopted by the DoD. For example, ASTM will provide DoD with one free copy of any document and grants royalty-free license to reproduce for DoD coordination review purposes only.

Non-Government standards can be referenced in military documents. Generally, referenced documents should be adopted to ensure availability of the specific issue reference.

The Adoption Process

The DoD has adopted many ASTM and SAE documents and uses adopted documents interchangeably with military documents. In the area of NDT, approximately one third of the 90 documents listed in DoD standardization AREA NDTI are non-Government documents (NGS). Area NDTI comprises all DoD specifications, standards, and handbooks and adopted NGS that deal with NDT methodology. The Army Materials Technology Laboratory (MTL) is the lead standardization activity for this area. I am the DoD liaison representative to ASTM E7, which is responsible for writing NDT method documents. ASTM E7 documents are referenced in hundreds of military and ASTM documents that cover various applications.

The DoD, National Institute of Standards and Technology, and the Naval Surface Warfare Center have been working with non-Government groups in the development of standards of mutual interest for many years. They have assumed leadership positions such as committee and subcommittee chairmenships on many non-Government committees and have aided in the development of documents with proper requirements and in their opinions satisfactory to the DoD. However, they are not in a position to ensure DoD approval of the documents. It is necessary to work within the DoD standardization system to ensure that there is a definite need for the document and that the documents are written in an acceptable manner. To ensure the adoption of a non-Government document, it is necessary to coordinate non-Government drafts beginning with the first draft of a proposed document through the proper DoD channels. To accomplish this, the DoD representative identifies the DoD standardization activities responsible for reviewing the particular non-Government document within the DoD. DoD representatives also contact DoD technical personnel knowledgeable on the particular document who may be missed by the standardization offices. It is important that one coordinated reply be sent to the NGS body.

Speeding Up Adoption by DoD-NGS Activities Working Together

DoD and non-Government bodies working in unison on projects independent of who started the task is the key to faster acceptance of the product of the work of both the Government and the non-Government bodies. An example of a non-Government initiated project is a General Electric radioscopy proposal which will be worked on by DoD and Industry and finally published as a military standard. This document will supply requirements for the "practice section" of ASTM Practice for Radioscopic Real Time Inspection (E 1255) and hopefully will be incorporated into that document (see radioscopy paragraph of this paper). For non-Government bodies to participate in Government-initiated projects, they must have knowledge of what is happening in DoD standardization. This is now accomplished by DoD standardization personnel participating in non-Government activities and presenting talks on DoD activities.

It is important to encourage non-Government bodies to become involved in DOD-initiated projects as many of the DOD projects end up with publication of non-Government documents. An example is ASTM Test Method for Primary Calibration of Acoustic Emmission Sensors (E 1106–86). This was a DoD project with the bulk of the work performed by the National Institute of Standards and Technology (NIST), formerly the National Bureau of Standards, under contract to MTL but with heavy input from the ASTM E07.04 subcommittee. The final product was an adopted ASTM document. Other examples are projects for reference radiographs for aluminum welds, titanium castings, and thick-wall aluminum castings. These projects were all seeded with DOD start-up money, have heavy DoD and ASTM input, and will be published as adopted ASTM documents. The titanium reference radiograph document was recently published.

The question arises, how can non-Government standardization bodies become aware of DoD-NDT projects in progress beyond the means previously described? One approach is by using the NDTI Program Plan. This document contains compilation of projects comprising most of the major ongoing NDT standardization tasks planned by DoD. Updates on tasks in the Program Plan are presented periodically at various DoD, ASTM, and ASNT meetings. The last updates were at the Fall 1989 ASNT Conference and the JANNAF NDT meeting, April 1990. A presentation was given to the DoD community in November 1990.

The DoD-NDT Program Plan

The purpose of this plan is to define the coordinated management program for standardization effort in the Nondestructive Testing and Inspection Area (NDTI). The Plan reflects agreement and commitment by the military services in the accomplishment of specific tasks within scheduled milestones. The Plan is the principal source of management information required for decision making at all levels within the DoD.

Revision 4 of the Plan was approved October 1989. It contains the standardization tasks and thoughts projected for the next two years. Many of the 18 tasks listed involve NGS writing bodies to some extent. The most used military documents include MIL-STD-1949A, Magnetic Particle Inspection; MIL-STD-6866, Penetrant Inspection (which project was completed in the last program plan); MIL-STD-410, Qualification and Certification of NDTI Personnel; MIL-STD-453, Radiographic Inspection; MIL-I-2154, Inspection, Ultrasonic, Wrought Metals, Process For; and MIL-I-6870, Inspection Requirements, Nondestructive for Aircraft and Missile Materials and Parts (Magnetic Particle, Penetrants, Radiography, Ultrasonic, Eddy Current). MIL-STD-271F(SH) Nondestructive Testing Requirements is intended for shipyard use but enjoys wider usage. MIL-STD-1949, MIL-STD-6866, MIL-STD-453, and MIL-STD-

2154 have already been put into ASTM format and balloted. DoD has no objection to cancelling its documents and superseding them with non-Government standards as long as the DoD requirements are carried over into the NGS. Once the NGS is published by its organization, it is then sent for DoD consideration to ensure acceptability. As already stated, initial and subsequent drafts of the NGS should have been coordinated with the DoD.

The following tasks of the current Program Plan are outlined and comments are welcome. DoD invites your interest in our tasks.

MIL-STD-410 Personnel Qualification

Proposed MIL-STD-410E specifies the qualification and certification requirements for nondestructive testing/nondestructive inspection personnel. Previous revisions of this specification addressed the requirements for personnel using penetrant, magnetic particle, ultrasonic, eddy current, and radiographic nondestructive testing/nondestructive inspection methods. This revision adds detailed requirements for acoustic emission and neutron radiographic methods as well as general requirements for any other nondestructive method for determining the acceptability of a product. In addition, this revision upgrades the designation of Level I, eliminates the Level I Special, adds an instructor level of qualification, and adds a recertification requirement for Level III. This document was published in January 1991.

Visual Acuity Requirements for Radiographers

To establish visual acuity test targets (transparencies) for measurement of visual acuity. Currently six sets of 50 transparencies have been produced. A protocol has been developed that will be used with the transparencies. A round-robin evaluation indicates that some images may be too readily detected and not provide adequate discrimination. There is a great deal of interest in this project. The transparencies could have carryover into other NDT methods.

Secondary Acoustic Emission Transducer Calibration

A technical outline of one possible secondary calibration method has been developed and distributed to ASTM E07.04.02. Development of a second method involving transient methods and data processing continues. Outlines of two possible secondary calibration methods were presented and discussed at the January 1988 meeting of ASTM E07.04.02 on AE sensors. Subsequently, considerable work was done on a laboratory prototype calibration setup using a plate 30 by 36 by 1-1/4 in. (76.2 by 91.44 by 3.175 mm) and the results compared with the results from the primary calibration. There are significant differences in the results attributable to the large [approximately 3.4 in. (1.90 mm)] diameter aperture of the test (commercial) transducer.

An understanding of the origins of the differences is being developed. The latest results using a steel transfer block approximately 16-1/4 by 16-1/4 by 7-1/2 in. (41.275 by 41.275 by 19.05 mm) appear to be very promising if a breaking glass capillary is used as the input and if the voltage-time waveform is truncated before the arrival of boundary reflections. Some additional effort will be required to complete this understanding and an acceptable secondary calibration method; however, significant progress has been made and it would seem all major technical problems are resolved. A draft document could begin in early January 1992.

MIL-STD-1907—Penetrant and Magnetic Particle Inspection, Soundness Requirements for Materials, Parts and Weldments

This standard was initially published with an incorrect number (MIL-STD-350). MIL-STD-1907 was published 7 Sept 1989. A Notice 1 published 22 March 1990 corrects a typographical error.

Eddy Current Coil Characterization

A military standard is being developed by NIST, Boulder, CO, which will be suitable for use in the field for calibrating eddy current coils. NIST/DOD has established that military depot and field level organizations are not the appropriate places to make eddy current probe characterization measurements. Presently, we are rewriting a draft military standard to require probe manufacturers to supply characterization measurements with the probes. The consensus of eddy current users with whom we have spoken is that this is a reasonable approach. It was pointed out that manufacturers of ultrasonic transducers are already required to provide these kinds of measurements on their products. Our revision of the draft standard will reflect a slightly different approach to the test method in that we will use a slotted block to produce the impedance probe response and we will require a d-c resistance measurement of the probe itself that can be verified by the procuring activity. We will be sending the revised document out for coordination when the changes have been completed.

MIL-STD-271, Nondestructive Testing Requirements for Metals

Since this document is intended for use on Navy ships and has been specifically designed for that purpose, it is not feasible to replace it with a non-Government document. Instead, the current approach is to create a table of non-Government documents that can be used alternatively for particular sections of MIL-STD-271.

Currently, non-Government documents are being evaluated by the NAVSEA NDE working group to determine which are acceptable to use as alternatives. Some documents that are being evaluated include the American Bureau of Shipyards (ABS) Rules for Nondestructive Inspection of Hull Welds, ASME Section V, ASTM Practice for Liquid Penetrant Inspection Method (E 165), ASTM Practice for Magnetic Particle Examination (E 709), and ASTM Guide for Radiographic Testing (E 94).

MIL-STD-XX40 Nondestructive Testing Acceptance Criteria

This standard will combine NAVSEA 0900-003-8000 "Visual, PT, MT Acceptance Standards" and applicable portions of NAVSEA 0900-LP-006-3010 "UT Inspection Procedure and Acceptance Standards for Welds," and NAVSEA 0900-003-9000 "Radiographic Acceptance Standards" to provide a single document containing acceptance standards.

Since this document is the summation of acceptance criteria for many NAVSEA documents, it is unlikely that it can be replaced by a non-Government document. However, the current approach is to create a table of non-Government documents that can be used alternatively for particular sections of MIL-STD-XX40. Some documents that will be evaluated for use as alternatives include the ABS Rules for Nondestructive Inspection of Hull Welds, ASME Section V and ASTM Reference Photographs for Liquid Penetrant Inspection (E 433).

MIL-I-6870, Inspection Program Requirements, Nondestructive for Aircraft and Missile, Materials and Parts

The Air Force has written a contract for the revision of this document. Due to the extensive work involved, the contract was extended. A draft document is due early 1991.

MIL-STD-453C Inspection Radiographic

A draft version of MIL-STD-453D, dated July 1990, was circulated August 1990 for a three-month coordination. In the General and Detail Requirements Section, there are numerous paragraphs with changes from MIL-STD-453C. The draft also contains information on storage of radiographs and reproduction of radiographs. An Appendix A, which was first included in a Notice 1 to MIL-STD-453C for qualification of equipment operations and procedures, is now part of the new draft. The Appendix A (for Army use) contains data item descriptions (DIDs) which ensure that the necessary parameters for performing radiographic testing are correctly noted. Comments are presently being studied.

MIL-STD-2175, Castings, Classification and Inspection

The military standard is being reviewed so that industry can readily comply with the requirements set therein. MIL-STD-1907 has been added. ASTM Reference Radiographs for Inspection of Aluminum and Magnesium Die Castings (E 505) will be added to account for die castings. The initial draft has just completed coordination and the comments are being studied.

Standard Guide for the Application Specific Selection of Acoustic Emission Sensors MIL-HDBK-788—published 25 JULY 89

The use of acoustic emission (AE) testing on major weapons systems is growing rapidly. However, each class of applications places its own demands on the AE system, particularly on the choice of sensors. Dozens of different sensor types are available and have been designed to maximize particular attributes. This handbook serves as a guide to assist the practitioner in selecting sensors most appropriate for a particular application.

MIL-STD-1949B, Magnetic Particle Inspection

A draft of MIL-STD-1949B was circulated in August 1990 to both government and industry users. Changes in the document include a test outlined in AMS 2641 for determination of wet particle contamination and sections on internal part inspection. Modifications from this draft will be reflected in the ASTM version of this document. (See new activities section of this paper for magnetic particle inspection using shims.)

The DoD is working closely with ASTM in writing an ASTM version of the military document. The approach is to develop both a tutorial document (completely revised E 709) and a requirements-driven document (MIL-STD-1949B in ASTM format basically). The magnetic particle inspection area presents an ideal situation in which DoD and ASTM are working together for the betterment of the NDT community.

MIL-HDBK-728, Nondestructive Testing

Replaces many military NDT handbooks. This handbook can easily be revised. Defense Logistics Agency may use it as an instructional tool for training purposes. The document is now being updated. Section 3 on magnetic particle inspection will be coordinated by July 1992.

MIL-STD-2154, Inspection, Ultrasonic

The electronic equipment requirements table will be changed to allow for better defined and more realistic equipment qualification requirements. This document has been formatted for ASTM ballot. If acceptable by the ASTM community, the military standard will be cancelled. A draft revision was coordinated in the fourth quarter of FY 90. Presently the document is being worked on in ASTM Committee E7.06.

ASTM EXXXX, Radiographic Inspection for Soundness of Welds in Aluminum by Comparison with Graded ASTM Reference Radiographs

This document is being developed in conjunction with ASTM E7.02. At the June 1990 ASTM meeting, data were presented on the preparation of aluminum weld reference radiographs. The Subcommittee reviewed the prototype radiographs of coarse and fine porosity in thick [1/2-in.(24.13 mm)] aluminum plate.

MIL-STD-XXX, Radioscopy

Draft 5 of a military standard is in coordination with DoD and ASTM E7.02 members. This draft essentially replaces Paragraph 5.2 of ASTM E 1255, which details the general practice. The purpose of this proposed military standard is to prescribe the radioscopic inspection requirements for all materials. Using ASTM E 1255 as the reference document, qualification of all aspects of radioscopic systems and the day-to-day control of their operation are itemized. This standard may allow radioscopic inspection, when determined by the Level III radiographic inspector of the contracting agency, to:

1. Replace radiographic inspection in existing applications with radioscopic methods, when equal or better inspection can be obtained.
2. Use radioscopy in new applications, where the contractually agreed quality level criteria can be met.

MIL-STD-2195(SH), Inspection and Detection of Measurement of Dealloying Corrosion on Aluminum Bronze and Nickel-Aluminum Bronze Components

MIL-STD-2195(SH) was published 28 April 1989.

Eddy Current Inspection of Heat Exchanger Tubing on Ships of the U.S. Navy

This document contains NDE requirements for eddy current in situ inspection of condenser and heat exchanger tubing. The proposed document will replace NAVSEA 0905-475-3010 and incorporate the latest techniques, technology, and equipment currently available for eddy current tubing inspection. An ASTM standard practice is available for this inspection [ASTM Practice for In-Situ Electromagnetic (Eddy Current) Examination of Nonmagnetic Heat Exchanger Tubes (E 690)]. However, the ASTM standard is very general and primarily intended as "guidance" document. At this time, it is not feasible to use E 690 for Navy applications.

As long-term effort, the ASTM committee will be requested to consider revising ASTM E

690 (or generate a new ASTM standard) to cover these inspections. However, it is recognized that this may not be feasible due to the specific Navy requirements in the document.

New Activities

Shims for Magnetic Particle Inspections

A reliable method for inspection system verification, according to MIL-STD-1949A, Magnetic Particle Inspection, is to use representative test parts containing defects of the type, location, and size specified in the acceptance requirements. When actual production parts are not available, then fabricated test parts with artificial defects must be used. Many people have asked whether shims can be used instead. There are many shims such as the Pie Gauge, Burmah Castrol Strips, and the QQI shims. Currently the DoD answer is "no."

A new study by NIST/DoD carried out by Lydon Swartzendruber of NIST will attempt to further answer the question. He has prepared a rough draft of a Standard Practice for Magnetic Particle Examination Using Shims. This draft document covers the use of shims to qualify and to verify magnetic particle test procedures. It is applicable only to the continuous method of magnetic particle inspection. It is especially recommended for use with the wet continuous method and when multidirectional magnetization is being used. This draft will be discussed at the January 1991 ASTM meeting.

Naval Sea Systems Command (NAVSEA)

NAVSEA has always had service-peculiar documents such as NAVSEA 0900-LP-003-8000, Surface Inspection Acceptance Standards For Metals. These documents did not appear in the DODISS and unless one had special involvement with NAVSEA, did not know of their existance.

Presently NAVSEA is incorporating requirements from several service-peculiar documents into one easy-to-obtain military standard. By doing this, the knowledge contained in these difficult-to-obtain documents will become available to all NDT users.

JANNAF Nondestructive Evaluation Subcommittee

The Joint Army-Navy-NASA-Air Force (JANNAF) Interagency Propulsion Committee is comprised of representatives from the Department of Defense (DoD) military services and the National Aeronautics and Space Administration (NASA) and is referred to as JANNAF. The purpose of JANNAF is to effect coordination and solution of propulsion problems and to promote the exchange of technical information in the field of missile, space, and gun propulsion technology based upon chemical or electrical energy release. My involvement is in the JANNAF Nondestructive Evaluation Subcommittee. This Subcommittee is comprised of five panels. They consist of the Advanced Inspection and Implementation, Solid Propulsion Unique Issues, Liquid Propulsion Unique Issues, Space Systems Issues, and the Component Inspection Standards Panels.

I am co-chairman of the Component Inspection Standards Panel. The Panel's task in general is to improve propulsion systems reliability by developing NDE standards and protocols which will include standardized calibration, inspection, and data analysis procedures, and common terminology. More specifically, the panel is now involved with developing a database of all NDT standards.

Development of the database is proceeding nicely thanks to the efforts of Bill St. Cyr from NASA Stennis Space Center, Mississippi. Bill has compiled an all-inclusive dBase 3 listing which includes nearly all aspects of NDE information such as standards, practices, procedures, books, handbooks, etc. Included in the database are DoD, API, ASME, ASTM, SAE, AIA (NAS document), ASNT, ASM, and AWS documents. Listed information includes area (MPI, PT, ET, etc.), source (DoD, ASTM, etc.), type (book, standard, etc.), document number, revision date, title and abstract, and scope or contents. The information is on two diskettes. Future plans include having the Nondestructive Testing Information Analysis Center (NTIAC) become the source for the diskette. NTIAC is operated for the U.S. Department of Defense by Texas Research Institute, Austin, Texas. Another option is to publish a military document with diskettes in the same fashion as an ASTM document with radiographs.

There is much interest in NDE inspection of adhesive joints. A document being studied is a state-of-the-art review on Nondestructive Evaluation of Adhesive Bond Quality by G. Light and H. Kwon, which is available from NTIAC.

A newly proposed project by John Moulder, Iowa State, explores electronic calibration of NDE inspection equipment, especially UT, RT, and ET probes and sensors. This involves identifying and promoting promising avenues of research that have potential applicability to electronically calibrated transducers which would provide NDE information that could also be used to determine material properties.

It is important that JANNAF information be interchanged with as large a body of technical persons as possible. JANNAF results could be published as ASTM documents. Persons participating in JANNAF meetings are usually different individuals than those who attend ASTM meetings. Most companies involved in JANNAF participate in NASA projects.

DoD's Initiatives in International Standardization

International standardization treaty and agreement documents may involve material and engineering practices. Examples of such documents are North Atlantic Treaty Organization Standardization Agreements (STANAGs), Quadripartite Army Standardization Agreements (QSTAGs), Quadripartite Navy Standardization Agreements (NAMSTAGs), and the Air Standardization Coordinating Committee (SCC) Air Standards (as produced by the Armies, Navies and Air Forces of United States, United Kingdom, Canada, Australia, including New Zealand in the case of ASCC Air Standards).

My knowledge of these groups is limited to participation on an American, British, Canadian and Australian (ABCA) armies team whose mission includes achieving the highest possible degree of interoperability through standardization of nondestructive testing techniques. The group is called an ABCA Quadripartite Working Group on Proofing, Inspection and Quality Assurance (QWG/PIQA). Its function is to agree that the methods that are followed during the manufacture and maintenance of material under the design control of each army conform to the accepted standard of that army. Its purpose is not to write methodology. An important point is that the standards agreed to are interoperable and mutually acceptable in many applications and that they form a body of technical information which should be available to technical organizations engaged in nondestructive testing in each army. The ABCA armies further agree to consult and, whenever possible, reach mutual agreement before introducing changes to any of their documents.

The NDT section of QWG/PIQA is actively preparing documents called QSTAGS, which are Quadripartite Standardization Agreements. QSTAGs are formal agreements between two or more armies defining the standardization achieved and to be maintained. The agreements are reviewed by armies for currency and validity on a continuing basis. The U.S. Army is actively involved in developing four QSTAGs. QSTAG 933, Calibrations of Ultrasonic Test Equipment by Means Other Than Test Blocks, uses ASTM E 1324, Guide For Measuring Some Electronic Characteristics of Ultrasonic Examination Instruments, as the acceptable

U.S. document. Another task, QSJAG 724 on Calibration of Ultrasonic Test Equipment With Standard Test Blocks, uses ASTM Practice for Evaluating Performance Characteristics of Ultrasonic Pulse-Echo Testing Systems Without the Use of Electronic Measurement Instruments (E 317) as the U.S.-accepted document. Other projects involve developing QSTAGS on conductivity measurement and coating thickness measurement. The conductivity QSTAG will involve MIL-STD-1537B, Electrical Conductivity Test For Measurement of Heat Treatment of Aluminum Alloys, Eddy Current Method, as the acceptable U.S. standard. QSTAG 938 on Coating Thickness Measurement involves ASTM E 376, "Standard Practice For Measuring Coating Thickness By Magnetic Field or Eddy-Current (Electromagnetic) Test Methods."

The Canadian Army is involved with the writing of QSTAGs for certification requirements for nondestructive testing personnel. The following QSTAGs have been developed and ratified for individual nondestructive testing methods: MPI (QSTAG 612), LPI (QSTAG 771), Radiography (QSTAG 274), Ultrasonics, (QSTAG 335), and Eddy Current (QSTAG 937).

Inactive projects include calibration of ultrasonic equipment and magnetic particle inspection.

Summary

The purpose of this paper has been to describe how DoD does business in the area of adoption, to describe on-going tasks in the Standardization Program Plan for NDT, and to invite the NDT community to participate in the tasks. The end product of these tasks can and usually is a NGS.

A brief overview was given of DoD's efforts on a JANNAF (Joint Army, Navy, NASA, and Air Force) Subcommittee for NDE. Projects were outlined, and a highly successful task of building a database of NDE documents was described.

Projects in which the U.S. Army is involved in an ABCA (American, Britain, Canada, and Australia) international standardization group are described. This effort involves not the writing standards but using or modifying standards so that they are interoperable and mutually accepted.

Jack C. Spanner, Sr.[1]

NDT Standards and the ASME Code

REFERENCE: Spanner, J. C., Sr., "**NDT Standards and the ASME Code,**" *Nondestructive Testing Standards—Present and Future, ASTM STP 1151,* H. Berger and L. Mordfin, Eds., American Society for Testing and Materials, Philadelphia, 1992, pp. 136–152.

ABSTRACT: Nondestructive testing (NDT) requirements and standards are an important part of the ASME Boiler and Pressure Vessel Code. The evolution of these requirements and standards is reviewed in this paper in the context of the unique technical and legal stature of the ASME Code. The coherent and consistent manner by which the ASME Code rules are organized is described, and the interrelationship between the various ASME Code sections, the piping codes, and the ASTM standards is discussed. Significant changes occurred in ASME Sections V and XI during the 1980s; these are highlighted along with projections and comments regarding future trends and changes in these important documents.

KEY WORDS: nondestructive testing (NDT), nondestructive examination (NDE), NDT standards, ASME Code, ASTM standards

Nondestructive testing (NDT) requirements and standards are an important part of the American Society of Mechanical Engineers (ASME) Boiler and Pressure Vessel Code (hereafter referred to as the ASME Code). The organizational structure of the American Society of Mechanical Engineers provides a great deal of autonomy for its Codes, Standards, and Accreditation and Certification Program. It also provides the advantages inherent in a professional society organized into committees, boards, and councils that enhance the advantages resulting from the multifaceted purposes of the society [1]. Quoting from C2.1.2 of the ASME Constitution: "The Society may approve or adopt any report, standard, code, formula, or recommended practice, but shall forbid and oppose the use of its name and proprietary symbols in any commercial work or business, except to indicate conformity with its standards or recommended practice." The technical stature of the ASME Code is generally regarded as being "in a class by itself" and is held in very high regard throughout most of the industrial world.

The ASME Code also enjoys a rather unique legal status by virtue of having been adopted as federal law by the Nuclear Regulatory Commission (NRC) [2], as well as by most of the 50 states in the United States and all of the provinces of Canada. Therefore, the NDT requirements published in the ASME Code truly have the force of law when applied within the jurisdictional limitations of the states that have adopted the ASME Code as law, the National Board of Boiler and Pressure Vessel Inspectors (National Board), and the NRC. The evolution of these requirements is reviewed herein in the context of the unique technical and legal stature of the ASME Code.

This paper also describes the interrelationships between the various sections of the ASME pressure vessel codes, the ASME piping codes, and the ASTM standards. A guided tour through Sections III, V, and XI, as well as B31.1 and B31.3, is used to introduce the reader to the format and content of these documents. This is followed by a discussion comparing the

---

[1] Staff engineer, Pacific Northwest Laboratory, Operated by Battelle Memorial Institute, Mail Stop K2-05, P.O. Box 999, Richland, WA 99352.

NDT requirements in Sections I, III, V, VIII, and XI. The term "nondestructive examination" (NDE) will be used rather than the terms "NDT" or "NDI" (nondestructive inspection). This is done for consistency with ASME Code usage where the term *examination* has a specific, different meaning than the terms *testing* and *inspection*.

Some of the more significant changes that occurred in ASME Sections V and XI during the 1980s are highlighted along with projections and comments regarding future trends and possible changes in these two important documents.

Interrelationships between the Various ASME Code Sections, the Piping Codes, and the ASTM Standards

The essential aspects of any NDT examination include the type, extent, and time of examination; the NDT techniques (procedures); acceptance standards/criteria; and the requirements for reporting results and repairs. Sections I, II, III, VIII, IX, and X cover all these essential NDT aspects, either directly or by reference. Section V covers only NDT methodology, procedures, and interpretation aids. Section XI covers all essential NDT aspects (including some methodology and procedural requirements) and also refers to Section V extensively for additional NDT methodology and procedural requirements.

The Section V rules on NDE methods and procedures are Code requirements only to the extent that they are specifically referenced and/or required by the other Code sections. Thus, the referencing Code sections (e.g., I, II, III, VIII, XI, etc.) specify the type, extent, and time of the NDE, as well as the applicable acceptance criteria/standards. Similarly, the ASME piping codes (e.g., B31.1 on Power Piping, B31.3 on Chemical Plant and Petroleum Refinery Piping, etc.) specify key parameters associated with NDE (i.e., type, extent, etc.), and also reference Section V for NDE methodology and procedural requirements.

Recognizing the stature and credibility of the American Society for Testing and Materials (ASTM) organization and its standards development process, ASME has referenced numerous ASTM standards pertaining to NDT/NDE. These documents, many of which were developed as consensus standards by the E-7 Committee on Nondestructive Testing, have been adopted by Section V.

Format and Contents

Section V

The contents of Section V only become ASME Code requirements when referenced by another Code section. Article 1 contains general requirements describing the responsibilities of the manufacturer and the authorized inspection agency, NDE procedures, and qualification of personnel. The ASME Code uses very explicit definitions for the terms "examination," "inspection," and "testing." Examination denotes the process of applying NDT methods, inspection refers to the functions performed by the "Authorized Inspector," and testing refers exclusively to pressure tests. The ASME Code (particularly Section XI) classifies all NDE methods into the three categories of volumetric, surface, and visual examinations.

Section V is organized in four major parts as follows: Subsection A entitled "Nondestructive Methods of Examination," Subsection B entitled "Documents Adopted by Section V," Appendix A entitled "Standard Definitions of Terms," and Appendix B entitled "Preparation of Technical Inquiries to the Boiler and Pressure Vessel Committee." Subsection A (Articles 2–12) provides specific, detailed requirements for all of the NDT methods whose use is specified in the other ASME Code sections. Subsection B (Articles 22–28) contains the complete text of the ASTM standards that have been adopted by, and referenced in, Section V.

Many ASTM "E" standards (e.g., E-165, E-214, etc.) have been adopted by the ASME Section V Committee on Nondestructive Testing. Documents designated by SA (e.g., SA-578, SA-388, etc.) use the ASTM designation "A" to refer to ferrous materials, whereas SB (SB-510, SB-513, etc.) use the ASTM designation "B" to refer to nonferrous materials.

Subsection A of Section V addresses each of the NDT methods separately and specifies the technique details and parameters that must be applied. The Subsection A articles include:

1. Article 2 on Radiographic (RT) Examination.
2. Article 3 on Radiographic (RT) Examination of Metallic Castings.
3. Article 4 on Ultrasonic (UT) Examination Methods for In-service Inspection.
4. Article 5 on Ultrasonic (UT) Examination Methods for Materials and Fabrication.
5. Article 6 on Liquid Penetrant (PT) Examination.
6. Article 7 on Magnetic Particle (MT) Examination.
7. Article 8 on Eddy Current (ET) Examination of Tubular Products.
8. Article 9 on Visual (VT) Examination.
9. Article 10 on Leak Testing (LT).
10. Article 11 on Acoustic Emission (AE) Examination of Fiber-Reinforced Plastic Vessels.
11. Article 12 on AE Examination of Metallic Vessels During Pressure Testing.

Subsection B entitled "Documents Adopted by Section V" includes the following: Article 22 on RT Standards, 23 on UT Standards, 24 on PT Standards, 25 on MT Standards, 26 on ET Standards, 27 on LT Standards, and 28 on VT Standards. These are all ASTM standards that have been adopted by, and are referenced and published in, ASME Section V. Their designation is changed slightly (for example, ASTM E-543 becomes ASME SE-543, ASTM A-388 becomes ASME SA-388, etc.), and a subheading is added to identify exceptions, modifications, limitations on applications, or it may state "Identical with ASTM specification . . . ."

The specific Section V Article(s) for each NDE method will next be reviewed separately to illustrate the interaction between the Subsections A and B documents with the referencing ASME Code section. These three sources must be utilized together to satisfy the ASME Code requirements for NDE of a given component or structure. In case of conflicts, the referencing ASME Code section supersedes Subsections A and B, and the methodology and procedural requirements in Subsection A supersede the SE, SA, SB, and SD document requirements.

For ASME Code applications, the volumetric NDE methods are RT, UT, ET, and AE; the surface examination methods are PT and MT; visual examination includes both direct and remote visual examination; and leak testing includes the four basic techniques applied during either pressure or vacuum testing.

Radiographic (RT) Examination—Article 2 specifies RT requirements for welds and materials (except castings), and Article 3 addresses RT of metallic castings. The ASTM standards referenced in Articles 2 and 3 are listed in Table 1 of this paper.

The documents SE-94, SE-142, and SE-242 define recommended practices for controlling RT quality during conventional film radiography, and SE-1255 describes radioscopic real-time examination. SE-1025 and SE-747 describe the use of hole-type and wire-type penetrameters, respectively. SE-999 is a standard guide for RT film processing, and SE-1079 describes densitometer calibration. SE-280 and SE-446 contain standard reference radiographs, and SE-586 is a glossary of standard RT terminology.

Article 2 specifies methodology and technique requirements for RT procedures, equipment and materials, calibration, examination, evaluation, and documentation. Appendix 1 is a mandatory appendix that specifies requirements for in-motion radiography, and Mandatory Appendix II (first published in the 1989 Addenda) provides requirements for real-time radioscopic examinations. The nonmandatory appendices include: (a) Technique Sketches for

TABLE 1—*Radiographic standards in Article 22.*

SE-94	Standard Practice for Radiographic Testing
SE-142	Standard Method for Controlling Quality of Radiographic Testing
SE-186	Standard Reference Radiographs for Heavy-Walled [2 to 4½ in. (51 to 114 mm)] Steel Castings
SE-242	Standard Reference Radiographs for Appearances of Radiographic Images as Certain Parameters Are Changed
SE-280	Standard Reference Radiographs for Heavy-Walled [4½ to 12 in. (114 to 305 mm)] Steel Castings
SE-446	Standard Reference Radiographs for Steel Castings Up to 2 in. (51 mm) in Thickness
SE-586	Standard Definitions of Terms Relating to Gamma and X-Radiography
SE-747	Standard Method for Controlling Quality of Radiographic Testing Using Wire Penetrameters
SE-999	Standard Guide for Controlling the Quality of Industrial Radiographic Film Processing
SE-1025	Standard Practice for Hole-Type Image Quality Indicators Used for Radiography
SE-1030	Standard Test Method for Radiographic Testing of Metallic Castings
SE-1079	Standard Practice for Calibration of Transmission Densitometers
SE-1255	Standard Practice for Radioscopic Real-Time Examination

Pipe or Tube Welds, (b) Equivalent IQI (Penetrameter) Sensitivity (EPS), and (c) Hole-Type Penetrameter Placement Sketches for Welds. Article 3 is organized similarly and contains requirements for procedures, equipment and materials, calibration, examination, evaluation, and documentation. The maximum X-ray voltages for steel, copper and/or high nickel alloys, and aluminum alloys are shown in figures, and the designation and selection of IQIs are specified in tables.

Ultrasonic (UT) Examination—The Section V articles on UT are Article 4 entitled "Ultrasonic Examination Methods for Inservice Inspection," Article 5 entitled "Ultrasonic Examination Methods for Materials and Fabrication," and Article 23 (supplementary ASTM standards). The supplementary UT standards referenced in Articles 4 and 5 are listed in Table 2.

Six of these are SA documents covering UT examination of various ferritic steel materials (i.e., forgings, plates, and castings). The four SB documents cover UT examination of nonferrous (i.e., nickel-alloy and aluminum-alloy) materials. The documents SE-114, SE-213, SE-214, SE-273, and SE-797 provide standard ASTM methodology requirements for UT, and SE-500 provides a glossary of definitions for UT. Article 4, first published in its present form during the 1980s, was developed to satisfy Section XI's need for methodology and procedural requirements during in-service inspection of nuclear power plant components. Article 5, first published as part of Appendix IX to Section III, describes or references requirements to be used in selecting and developing UT procedures for welds, parts, components, materials, and thickness determinations to be used in conjunction with Sections II, III, VIII, IX, and XI during the manufacture, construction, and/or ISI of pressure-retaining components built in accordance with the ASME Code. Thus, Article 4 is used exclusively for Section XI ISI applications, whereas Article 5 is used for all ASME Code applications requiring UT examination, including Section XI. In the scopes of both Articles 4 and 5, the statement is made that the referencing Code section shall be consulted for specific requirements on: (a) personnel qualification/certification, (b) procedures, (c) examination system characteristics, (d) retention and control of calibration blocks, (e) extent/volume of examination, (f) acceptance standards, and (g) records and reports.

Both Articles 4 and 5 define requirements for procedures and techniques, scanning processes, equipment and supplies, applications by product form and/or type of component, evaluation, and reports and records. Article 4 includes two mandatory appendices entitled "Screen

TABLE 2—*Ultrasonic standards in Article 23.*

SA-388	Recommended Practice for Ultrasonic Testing and Inspection of Heavy Steel Forgings
SA-435/SA-435M	Standard Specifications for Straight-Beam Ultrasonic Examination of Steel Plates
SA-577/SA-577M	Standard Specifications for Ultrasonic Angle-Beam Examination of Steel Plates
SA-578/SA-578M	Specifications for Straight-Beam Ultrasonic Examination of Plain and Clad Steel Plates for Special Applications
SA-609	Standard Specifications for Longitudinal Beam Ultrasonic Inspection of Carbon and Low-Alloy Steel Castings
SA-745	Standard Practice for Ultrasonic Examination of Austenitic Steel Forgings
SB-509	Specification for Supplementary Requirements for Nickel Alloy Plate for Nuclear Applications
SB-510	Specification for Supplementary Requirements for Nickel Alloy Rod and Bar for Nuclear Applications
SB-513	Specification for Supplementary Requirements for Nickel Alloy Seamless Pipe and Tube for Nuclear Applications
SB-548	Standard Method for Ultrasonic Inspection of Aluminum-Alloy Plate for Pressure Vessels
SE-114	Recommended Practice for Ultrasonic Pulse-Echo Straight-Beam Testing by the Contact Method
SE-213	Standard Practice for Ultrasonic Inspection of Metal Pipe and Tubing
SE-214	Standard Practice for Immersed Ultrasonic Examination by the Reflection Method Using Pulsed Longitudinal Waves
SE-273	Standard Practice for Ultrasonic Examination of Longitudinal Welded Pipe and Tubing
SE-797	Standard Practice for Thickness Measurements by Manual Contact Ultrasonic Method
SE-500	Standard Definitions of Terms Relating to Ultrasonic Testing

Height Linearity" and "Amplitude Control Linearity," and four nonmandatory appendices to define layout of vessel reference points, general techniques for angle beam calibrations, general techniques for straight beam calibrations, and data records for a planar reflector. Article 5 has two mandatory appendices on "Screen Height Linearity" and "Amplitude Control Linearity," and a nonmandatory appendix describing an alternative calibration block configuration.

The specific applications listed in Article 4 include vessels, pumps, and valves (including welds), and the specific applications listed in Article 5 include welds, cladding, and thickness measurements, as well as plate, forgings and bars, tubular products, and bolting materials.

Liquid Penetrant (PT) Examination—The liquid penetrant method is described in Articles 6 and 24 and is used extensively for ASME Code applications as a surface examination method. Liquid penetrants offer a very sensitive method for detecting surface discontinuities in both ferrous and nonferrous materials. Article 6 specifies PT requirements including methodology, technique, and procedural details. SE-270 is a glossary of standard PT definitions. The ASTM standards referenced in Article 6 are listed in Table 3.

TABLE 3—*Liquid penetrant standards in Article 24.*

SD-129	Standard Test Method for Sulfur in Petroleum Products (General Bomb Method)
SD-808	Standard Method of Test for Chlorine in New and Used Petroleum Products (Bomb Method)
SE-165	Standard Practice for Liquid Penetrant Inspection Method
SE-270	Standard Definitions of Terms Relating to Liquid Penetrant Inspection

TABLE 4—*Magnetic particle standards in Article 25.*

SE-269	Standard Definitions of Terms Relating to Magnetic Particle Inspection
SE-709	Standard Practice for Magnetic Particle Examination

The document SE-165 provides a detailed description/discussion of the two major liquid penetrant techniques (i.e., visible and fluorescent) and the three processes that are available (i.e., water-washable, post-emulsifiable, and solvent-removable). The documents SD-129 and SD-808 describe test methods for measuring impurities in penetrant materials (i.e., sulphur and chlorine, respectively). In SE-165, Annex A2 entitled "Methods for Measuring Total Chloride Content in Combustible Liquid Penetrant Materials" and Annex A3 entitled "Method for Measuring Total Fluoride Content in Combustible Liquid Penetrant Materials" were added in the mid-1980s in response to concerns arising from applications in the aerospace and nuclear industries regarding the potential for the impurities in PT materials to cause stress corrosion cracking and other adverse effects.

Article 6 contains requirements and methodology guidance for procedures, techniques, selection of penetrant materials, surface preparation, examination process, interpretation, evaluation, and procedures for nonstandard (either high or low) temperature applications.

Magnetic Particle (MT) Examination—The magnetic particle method is described in Articles 7 and 25. Article 7 specifies requirements for magnetic particle examination of materials and components. The MT method is effective for detecting surface and near-surface discontinuities in ferromagnetic materials. Sensitivity is greatest for surface discontinuities and decreases rapidly with increasing depth below the surface. The ASTM standards referenced in Article 7 are listed in Table 4.

The document SE-709 provides a detailed discussion/description of MT methodology and procedural details, as well as an overall tutorial discussion of the MT method. The document SE-269 is a glossary of standard MT definitions.

Article 7 provides methodology and technique requirements for procedures, examination techniques and materials, surface preparation, magnetization, demagnetization, equipment calibration, examination techniques and coverage, evaluation of indications, and reports.

Eddy Current (ET) Examination—The eddy current method of examination is described in Articles 8 and 26. ASME Code applications of the ET method are generally limited to the examination of tubular products such as steam generator, heat exchanger, and condenser tubes. Although the ASME Code specifies ET as a volumetric examination method, it should be recognized that the sensitivity is always greatest to discontinuities located at or near the surface closest to the ET probe and diminishes rapidly with increasing material thickness. The supplemental ASTM standards referenced in Article 8 are listed in Table 5.

TABLE 5—*Eddy current standards in Article 26.*

SE-215	Recommended Practice for Standardizing Equipment for Electromagnetic Testing of Seamless Aluminum-Alloy Tube
SE-243	Standard Practice for Electromagnetic (Eddy Current) Testing of Seamless Copper and Copper-Alloy Tubes
SE-268	Standard Definitions of Terms Relating to Electromagnetic Testing
SE-309	Standard Practice for Eddy-Current Examination of Steel Tubular Products Using Magnetic Saturation
SE-426	Recommended Practice for Electromagnetic (Eddy-Current) Testing of Seamless and Welded Tubular Products, Austenitic Stainless Steel and Similar Alloys
SE-571	Standard Practice for Electromagnetic (Eddy-Current) Examination of Nickel and Nickel Alloy Tubular Products

The documents SE-215, SE-243, SE-309, SE-426, and SE-571 all describe methodology and techniques for examining tubes and/or tubular products, and SE-243 is a glossary of standard terminology for the ET method.

Article 8 contains methodology and technique requirements for procedures, reference/calibration specimens, equipment calibration and qualification, and interpretation and evaluation of indications. Whereas Article 8, per se, is a relatively concise document, Appendix 1 entitled "Eddy Current Examination Method for Installed Nonferromagnetic Steam Generator Heat Exchanger Tubing" provides a detailed discussion/description of ET methodology, procedures, and techniques for single frequency equipment.

Visual (VT) Examination—Article 9 describes methodology and technique requirements for the visual examination method. For most ASME Code applications, the VT method is used in conjunction with fabrication, hydrostatic testing, leak testing, etc. Visual interpretation and evaluation of discontinuities detected using the other NDE methods is considered to be outside the scope of Article 9. For ASME Code applications, the VT method is generally used to evaluate surface conditions, alignment of mating surfaces, shape, or evidence of leakage. In addition, the VT method is used to evaluate the subsurface condition of translucent composite materials. Article 9 includes a description of the VT methodology in addition to specific requirements for written procedures, applications (i.e., direct visual, remote visual, and translucent visual examination), evaluation of results, and reports and records. The single ASTM standard referenced in Article 9 is listed in Table 6.

Leak Testing (LT)—Article 10 describes methodology and requirements for conducting leak testing (LT), and three ASTM standards are listed in Table 7. The documents SE-432 and SE-479 are recommended guides for selecting techniques and preparing specifications, respectively, and SE-425 is a glossary of standard LT definitions.

Article 10 contains a concise summary of general requirements for the LT method and is supplemented by mandatory appendices containing detailed requirements for each of the six Code-acceptable LT techniques. Article 10 provides requirements for test article preparation, calibration procedures, evaluation of results, written procedures, and reports and records. The mandatory appendices provide general and specific requirements for the following techniques: (1) bubble test—direct pressure technique, (2) bubble test—vacuum box technique, (3) halogen diode detector probe tests, (4) helium mass spectrometer test—detector probe technique, (5) helium mass spectrometer test—tracer probe and hood techniques, and (6) pressure change test.

Acoustic Emission (AE) Examination—Articles 11 and 12 cover acoustic emission examination of plastic and metallic vessels, respectively. Article 11 is entitled "Acoustic Emission Examination of Fiber-Reinforced Plastic Vessels," and Article 12 is entitled "Acoustic Emission Examination of Metallic Vessels During Pressure Testing." Article 11 is applicable to both new and in-service fiber-reinforced plastic (FRP) vessels examined under pressure, vacuum, or other applied stress. Article 12 is applicable for AE examination of new metallic pressure vessels during acceptance pressure testing.

Both Articles 11 and 12 include general methodology and technique requirements for AE examination, plus specific requirements for equipment and supplies, applications, procedures, calibration, evaluation of results, and reports and records. Although at least nine ASTM stan-

TABLE 6—*Visual examination standard in Article 28.*

SD-2563	Specifications for Classifying Visual Defects in Glass-Reinforced Laminates and Parts Made Therefrom

TABLE 7—*Leak testing standards in Article 27.*

SE-425	Standard Definitions of Terms Relating to Leak Testing
SE-432	Standard Recommended Guide for the Selection of a Leak Testing Method
SE-479	Recommended Guide for Preparation of a Leak Testing Specification

dards have been published for the AE method, none have been adopted and printed as references in Section V, except the AE glossary document, SE-610.

The mandatory appendices to Article 11 include equipment performance requirements and instrument calibration requirements. A nonmandatory appendix provides sensor placement guidelines. The mandatory appendices to Article 12 include instrumentation performance requirements, instrument calibration, and cross-referencing. The nonmandatory appendices to Article 12 provide sensor placement guidelines and supplemental information for conducting AE examinations.

Articles 11 and 12 both state that discontinuities/relevant indications detected by AE shall be evaluated by other NDE methods. Articles 11 and 12 also contain specific requirements regarding written procedures, sensor placement, equipment calibration, and interpretation of AE response signals.

Personnel Qualification—The ASME Section V requirements state that all NDE personnel shall be qualified in accordance with the requirements of the referencing Code section. Qualification of personnel is usually in accordance with SNT-TC-1A or an alternate system specifically accepted by the referencing Code section. If Section V is not referenced, qualification may simply involve a demonstration that the personnel performing NDE are competent to do so in accordance with established procedures.

Section XI

Section XI of the ASME B&PV Code is entitled "Rules for Inservice Inspection of Nuclear Power Plant Components." This Code provides rules for the examination, testing, and inspection of components and systems in a nuclear power plant. Section XI does not become applicable until the requirements of the construction code (i.e., Section III) have been satisfied. The rules of Section XI constitute requirements to maintain the nuclear power plant and return it to service following outages in a safe and expeditious manner. Section XI specifies a mandatory program of examination, testing, and inspection to assure adequate safety.

The owner of a nuclear power plant is responsible for developing a program to demonstrate conformance with the requirements of Section XI. Section XI requires the services of an authorized nuclear inservice inspector (ANII) whose duties are to verify that the responsibilities of the owner and the mandatory requirements of Section XI are met.

Section XI rules and requirements specify, as a minimum, the responsibilities, areas subject to inspection, provisions for accessibility and inspectability, examination methods and procedures, procedure qualifications, frequency of inspection, record keeping and report requirements, procedures for evaluation of inspection results, disposition of results of evaluations, and repair requirements. Section XI also provides rules for design, fabrication, installation, and inspection of replacement components.

Section XI consists of three divisions as follows: Division 1—Rules for Inspection and Testing of Components of Light-Water Cooled Plants, Division 2—Rules for Inspection and Testing of Components of Gas-Cooled Plants, and Division 3—Rules for Inspection and Testing of Components of Liquid-Metal Cooled Plants.

The three types of NDE employed during inservice inspection are categorized as (1) visual, (2) surface, and (3) volumetric. Remotely controlled equipment is permitted because some NDE is required on irradiated components or in radiation zone areas. In general, NDE for Section XI applications must be performed in accordance with the methodology requirements of Section V (some of which are also specified in Section XI and some are imposed by direct reference to Section V). Many of these requirements are supplemented by methodology and procedural requirements that are explicitly defined in Section XI.

Visual examinations are classified as VT-1 (conducted to determine the condition of a part, component, or surface); VT-2 (conducted to locate evidence of leakage from pressure-retaining components); and VT-3 (conducted to assess the general mechanical and structural condition of components and supports).

Volumetric examinations are performed to detect discontinuities throughout the volume of a material or component and may be conducted from either the inside or outside surface of a component using RT, UT, or ET. Surface examinations are performed to detect the presence of surface-opening discontinuities and are performed using MT or PT.

Alternate examination methods, a combination of NDE methods, or newly developed techniques may be substituted for the methods specified in Section XI, provided the results are demonstrated to be equivalent or superior to those of the specified method.

The essential NDE aspects, as well as methodology and procedural requirements, for VT and UT examination are specified in Section XI. For UT these are found in mandatory appendices I (Ultrasonic Examinations), III (Ultrasonic Examination of Piping Systems), and VI (Ultrasonic Examination of Bolts and Studs). These three appendices, in turn, contain specific UT methodology and procedural requirements; both in these appendices and by direct reference to Articles 4 and 5 of Section V.

Methodology and procedural requirements for RT, PT, and MT are specified as direct references to Articles 2, 6, and 7, respectively, of Section V. Methodology and procedural requirements for ET (limited to examination of heat exchanger/steam generator tubing) are specified in Appendix IV, which includes a direct reference to Article 8 of Section V.

Personnel performing NDE under Section XI must be qualified and certified in accordance with SNT-TC-1A, plus various additional requirements, as specified in IWA-2300, IGA-2300, or IMA-2300, as applicable.

Significantly, six of the eight mandatory appendices to Division 1 (light-water reactor plants) are devoted to NDE topics and contain primarily technical requirements; the other two appendices are devoted to administrative topics.

Mandatory Appendix I specifies UT methodology and procedural requirements for various applications such as vessels, piping, and bolting materials. Appendix III specifies the methodology, equipment, and procedural requirements for UT of piping systems. Appendix VI provides detailed methodology and procedural requirements for UT of bolting materials. This appendix specifically addresses the qualification of personnel and procedures, as well as providing general and specific examination requirements.

The methodology and procedural requirements for ET examination of installed nonferromagnetic steam generator/heat exchanger tubing is defined in Appendix IV. This appendix refers to Article 8 (Appendix I) of Section V and also provides supplementary requirements covering personnel, procedures, calibration, records, evaluation, and the examination report.

Recent additions to Section XI include Appendix VII entitled "Qualification of Nondestructive Examination Personnel for Ultrasonic Examination" and Appendix VIII entitled "Performance Demonstration for Ultrasonic Examination Systems." Appendix VII specifies requirements for the training and qualification of UT/NDE personnel in preparation for employer certification to perform UT examination in accordance with Section XI. Appendix VIII specifies performance demonstration requirements for the personnel, equipment, and

TABLE 8—*Supplements to Appendix VIII of Section XI.*

Supplement 1	Evaluating Electronic Characteristics of Ultrasonic Instruments
Supplement 2	Qualification Requirements for Wrought Austenitic Piping Welds
Supplement 3	Qualification Requirements for Ferritic Piping Welds
Supplement 4	Qualification Requirements for the Clad/Base Metal Interface of Reactor Vessels
Supplement 5	Qualification Requirements for Nozzle Inside Radius Sections
Supplement 6	Qualification Requirements for Reactor Vessel Welds Other Than Clad/Base Metal Interface
Supplement 7	Qualification Requirements for Nozzle-to-Vessel Welds
Supplement 8	Qualification Requirements for Bolts and Studs

procedures that are used to detect and size flaws in accordance with Section XI. These requirements apply to personnel who detect, record, or interpret indications, or size flaws in welds or components. Appendix VIII includes the specific application supplements that are listed in Table 8.

Section III on Nuclear Power Plant Components

Section III, Division 1, consists of six separate subsections plus a series of appendices. Although not obvious at first glance, the format is logical, coherent, and easy to use. NDE requirements are primarily found in three locations: the 2000 articles on materials, the 4000 articles on fabrication, and the 5000 articles on examination. Section III includes a choice of rules that provide three levels of structural integrity assurance (i.e., Classes 1, 2, and 3), plus additional rules for metal containment vessels, component supports, and core support structures. Subsection NCA contains QA requirements and general rules that are applicable to all classes of components. Specific rules for each class of service are found in the applicable subsection (e.g., NB for Class 1, NC for Class 2, etc.).

Weld examination requirements are included in NB-5000, NC-5000, ND-5000, etc. (i.e., NX-5000). Examination requirements for each product form are listed for each class of service (i.e., Class 1, Class 2, or Class 3). Major product forms include plate, forgings and bars, tubular products and fittings, castings, and bolting materials. The type, time, and extent of examination for materials, as well as the acceptance standards, are located in the NX-2500 paragraphs.

The NX-5000 Articles describe the requirements for examining fabricated components (both in-process and final examinations). The volumetric NDE methods are radiography and ultrasonics, and these are only occasionally considered to be equivalent (i.e., interchangeable) in Section III. The surface methods (MT and PT) are always considered to be interchangeable, when technically this is not necessarily justified in this author's opinion. For example, MT may detect subsurface discontinuities, whereas PT will not. Furthermore, PT may exhibit a magnification of ×5 or more for small discontinuities, whereas MT exhibits little or no magnification. The acceptance standards for Sections II, III, VIII, and IX are based on the size of the NDE indication, rather than on the size of the discontinuity that caused the indication. This is consistent with general industry practice and tends to be a conservative and pragmatic approach.

ASME Piping Codes

B31.1 Power Piping—The B31.1 Power Piping Code prescribes minimum requirements for the design, materials, fabrication, erection, testing, and inspection of power and auxiliary service piping systems. Power piping systems covered by this Code apply to all piping components including flanges, fittings, etc., and miscellaneous appurtenances unless specifically excluded. This Code includes, but is not limited to, steam, water, oil, gas, and air services.

The terms "examination," "inspection," and "testing" are explicitly defined and used as they are in the ASME Boiler and Pressure Vessel Code. Examination denotes use of the NDT processes, inspection denotes verifying NDT and other activities, and testing refers to pressure or vacuum testing (generally as part of the acceptance process).

The B31.1 Code includes (in Chapter VI) requirements for visual examination, magnetic particle examination, liquid penetrant examination, radiography, leak testing, and personnel qualification. The MT, PT, and RT requirements specify that these examinations shall be performed in accordance with the appropriate ASME Section V Article; however, the VT and LT requirements do not include a reference to Section V. The type, time, and extent of examination are specified for each of the NDT methods covered by Chapter VI, along with explicit, narrative acceptance criteria. Generally, cracks and other linear indications are rejectable, and the acceptability of other flaw types is judged according to NDE indication size and/or weld thickness.

B31.3 Chemical Plant and Petroleum Refinery Piping—The B31.3 Chemical Plant and Petroleum Refinery Piping Code prescribes requirements for the materials, design, fabrication, assembly, erection, examination, inspection, and testing of piping. This Code applies to piping and piping system components used for transporting various fluids. The terms "examination," "inspection," and "testing" are defined and used the same as in other ASME codes. References to the "Inspector" are to the Owner's Inspector or the Inspector's delegate.

The B31.3 Code includes requirements for VT, MT, PT, RT, UT, and LT. B31.3 uses the term "random" to refer to UT and RT of not less than 5% of the total weld length for piping used in normal fluid service. The other two, more exacting, service conditions require more extensive examination, as might be expected. For specifying the extent of examination, B31.3 uses the terms 100% examination, random examination, spot examination, and random spot examination.

The acceptance criteria for welds are specified in a table that shows a matrix of flaw types as a function of examination method. The specific acceptance criteria are then categorized under three different types of fluid service. The type of examination for evaluating imperfections is also specified in a matrix showing the type of flaw as a function of examination method. B31.3 describes requirements for both random and 100% volumetric examination using either RT or UT.

The B31.3 Code specifies that VT, MT, PT, RT, and UT be performed in accordance with the appropriate requirements of Section V, and that the NDE personnel must have training and experience commensurate with the needs of the specified examination. (B31.3 suggests only that the ASNT document SNT-TC-1A be used as a guide.) Specific guidelines for performing different types of leak tests are specified except that Article 10 of ASME Section V is referenced for gas and bubble formation testing. Cracks and lack of fusion are generally unacceptable, irregardless of length, and the acceptability of other flaw types is judged according to estimated flaw indication size as a function of weld thickness.

ASTM Standards

Since this paper was prepared for an ASTM symposium, and many of the other papers address various aspects of the ASTM standards available for NDE applications, this subject will receive limited emphasis. However, it seems appropriate to note that ASTM was ". . . formed for the development of standards on characteristics and performance of materials, products, systems, and services; and the promotion of related knowledge. (In ASTM terminology, standards include test methods, definitions, recommended practices, classifications, and specifications.)" [*3*] Also "An ASTM standard represents a common viewpoint of those parties concerned with its provisions, namely producers, users, consumers, and general interest groups. It is intended to aid industry, government agencies, and the general public. The use of an ASTM standard is purely voluntary. . . . Because ASTM standards are subject to periodic review and revision, those who use them are cautioned to obtain the latest revision." [*4*]

One key phrase is that "The use of an ASTM standard is purely voluntary." However, once invoked by a referencing standard such as the ASME Code, or via formal agreement between a producer and user, the requirements of an ASTM standard may, and often do, become contractually binding.

ASME Code Cases and NRC Requirements

Code cases are permissive and may be used beginning with the date of approval by the ASME Council. Generally, only Code cases that are specifically identified as pertaining to the rules of a particular Code section may be used with that section. Code cases are also generally only used by mutual consent of the plant owner and the relevant holder of a valid ASME Certificate of Authorization. In some situations, the designer must also consent to this agreement.

Prior to use of an ASME-approved Code edition or Code case for an application governed by ASME Section XI and the NRC regulations, the Code edition and/or Code case must be formally reviewed and accepted by the NRC staff. The Federal Code of Regulations (10 CFR 50) lists the Code editions that have been formally accepted by the NRC, and NRC Regulatory Guide 1.147 entitled "Inservice Inspection Code Case Acceptability—ASME Section XI, Division 1" lists those Section XI ASME Code cases that are generally acceptable to the NRC staff for implementation in the inservice inspection of light-water-cooled nuclear power plants. Regulatory Guide 1.147 is periodically updated to accommodate new Code cases and revisions of existing Code cases. Endorsement of a Code case by Regulatory Guide 1.147 constitutes acceptance of specific technical requirements for applications not precluded by regulatory authority or other requirements. Some Code cases are accepted in total as issued by ASME, while others are accepted subject to technical and/or administrative restrictions and limitations.

Comparison—NDE Requirements in Section XI Versus Other ASME Codes

The bases for the acceptance standards specified in Section XI differ considerably from the approach used in all the other Code sections (including the piping codes) that specify acceptance standards/criteria. First, the Section XI acceptance standards are based on the actual size of the flaw (as estimated using NDE or otherwise), whereas the other Code sections base their acceptance criteria on the size of the NDE indication caused by the flaw (which may be quite different than the size of the actual flaw). Another obvious difference is that the evaluation processes used to determine the rejection criteria in Section XI are based on stress analysis and fracture mechanics calculations, whereas the criteria used for determining the rejection criteria in all other Code sections (including the piping codes) are based on a combination of fabrication experience, workmanship criteria, and engineering judgement. Another significant difference between Sections I, III, VIII, and XI of the ASME Code is that Section XI puts primary dependence on the UT method for volumetric examinations, whereas Sections I, III, and VIII rely primarily on RT for volumetric examinations.

The differences outlined above, although technically significant, have generally caused fewer problems than might be expected. One reason for this is that beginning almost with the inception of the Section XI preservice requirements, most vessel manufacturers usually conducted a preservice examination using the Section XI rules and criteria (i.e., UT) before the component left the shop, even though such an examination was not yet required by the applicable Code rules.

Differing requirements for the qualification and certification of NDE personnel are also evident when comparing the various Code sections. For example, the NDE/PQ requirements in Sections III and XI were very similar throughout most of the 1980s, and both were much more demanding than Sections I, V, or VIII. Sections III and XI are also more demanding regarding the requirement for a written procedure than Sections I, VIII, and the piping codes. The requirements of Section III regarding the extent of examination tend to be much more demanding than Section VIII, which permits spot radiography during vessel fabrication under certain circumstances.

Changes and Trends During the 1980s

Section V

During the 1980s, significant changes were made to the Code requirements for NDE in certain areas, whereas in other areas the Code requirements remained virtually unchanged. In general, the NDE requirements became more complex, more demanding, and more detailed. For example, the Section V Code book expanded from 330 pages to well over 500 pages. New articles were added to provide rules for acoustic emission (AE) examination of fiber-reinforced plastic vessels (Article 11) and metallic vessels during pressure testing (Article 12). These two articles alone account for an increase of about 40 pages.

In the radiography area, one of the penetrameter selection tables was eliminated, use of the term "image quality indicator (IQI)" increased as a replacement for the term penetrameter, rules for in-motion radiographic techniques were developed and published, Article 2 was completely reorganized, and a new Article 3 was developed to provide rules for RT examination of metallic castings. Following experimental studies that showed the value of the wire-type IQI, this type penetrameter was approved as an acceptable alternative to the traditional plaque-type penetrameters for both Article 2 and Article 3 applications. Another significant change to Article 2 was a mandatory appendix permitting the use of real-time radioscopic techniques for examining weldments.

Articles 4 and 5 were retitled "Ultrasonic Examination Methods for Inservice Inspection" and "Ultrasonic Examination Methods for Materials and Fabrication," respectively, and they both underwent major restructuring. Although numerous editorial changes were made to clarify Articles 4 and 5, surprisingly few changes to the basic technical and procedural rules were made during the 1980s. One of the few significant changes (which occurred late in the decade) added the requirement for a 70° beam to the existing requirements for 45 and 60° beams, and concurrently reduced the minimum recordable response amplitude from 50 to 20% of the reference level. These revisions to Article 4 were incorporated at the request of Section XI based on the results of international studies showing a compelling need for these changes. Numerous adjustments were made to various design details to improve the utility of the calibration blocks described in Articles 4 and 5. Rules for inservice inspection of vessel nozzles were, unfortunately, still not yet available as of issuance of the 1990 Addenda, although progress has been recently achieved in this area.

Article 6 on "Liquid Penetrant Examination" and Article 7 on "Magnetic Particle Examination" generally experienced no dramatic changes during the 1980s, as might be expected for these two methods. Significantly, the requirements for determining the contaminant content in liquid penetrant materials used on nickel-based alloys, austenitic stainless steels, and titanium were substantially expanded from the 1980 edition of Section V.

Article 8 on "Eddy Current Examination of Tubular Products," Article 9 on "Visual Examination," and Article 10 on "Leak Testing" also underwent major format changes for consistency with the balance of Section V, but there were few major changes in the technical requirements specified in these three Articles. Article 10 on leak testing was expanded to provide a much more comprehensive discussion of the techniques involved with leak testing. These expansions occurred through the use of Mandatory Appendices I through VI, which considerably strengthen and enhance the utility of Article 10. As noted earlier, Articles 11 and 12 were added to provide requirements for acoustic emission examination of plastic vessels and metal vessels, respectively.

Section XI

Among the more significant revisions to Section XI during the 1980s were new rules for the qualification and certification of NDE personnel. Dramatic changes occurred in the underlying philosophy and criteria, as well as in the specific requirements, for certifying NDE personnel. The major changes are reflected in Subarticle IWA-2300 "Qualifications of Nondestructive Examination Personnel," as supplemented by Appendix VII "Qualification of Nondestructive Examination Personnel for Ultrasonic Examination" and Appendix VIII "Performance Demonstration for Ultrasonic Examination Systems."

Significantly, the ASME Code (Sections III and XI) pioneered the change in recertification interval for Level III personnel from three years to five years. The corresponding change to ASNT's SNT-TC-1A document did not occur until some time later. Section XI formally accepted a valid ASNT Level III certificate for satisfying the Basic and Method Examinations requirements, although a separate, Specific Examination, that covers Section XI applications and references must still be administered.

Major changes in philosophy and requirements for qualifying UT personnel occurred with the issuance of Mandatory Appendices VI, VII, and VIII. Appendix VI specifies requirements for UT examination of bolts and studs and initiated the Section XI trend toward qualifying personnel and procedures using a performance demonstration test. The purpose of the performance demonstration is to evaluate the ability of personnel to operate the UT system and collect and interpret data as specified in the procedure. Satisfactory completion of the procedure qualification process may also serve to qualify personnel. A significant change is that the performance demonstration requires full-scale components with simulated actual flaws.

Issuance of Appendix VI was followed by Appendix VII, which specifies overall requirements for the training and qualification of UT personnel. This appendix includes expanded criteria for the qualification of trainees and Levels I, II, and III personnel. The responsibilities, as well as the qualification criteria, for an NDE Instructor are also included in this appendix. Appendix VIII reduces the prerequisite initial experience for qualifying Levels I and II personnel, increases the prerequisite experience requirement for Level III personnel, and eliminates the simultaneous experience provisions permitted by SNT-TC-1A. This appendix also imposes a periodic training requirement of 10 h per year for all NDE personnel. In addition, the basic outline for L-III training courses was tailored to emphasize nuclear applications. About half of the Specific Examination questions must now cover Section XI applications, and the "question bank" must contain at least twice the minimum number of questions per examination.

Appendix VIII, which describes a performance demonstration process for UT personnel, equipment, and procedures, was published in the 1989 Addenda to Section XI. This Appendix specifies general and specific UT system qualification requirements, tolerances on the essential variables, and documentation requirements.

On the basis of international studies on UT system performance and reliability (PISC-II studies), an additional angle beam examination at 70° was added to the previous Section XI (and Section V) requirements for beam angles of 45 and 60°. In addition, a requirement to investigate and record all angle beam reflectors that produce a response greater than 20% of the reference level was also imposed, whereas the previous level had been 50%. As noted earlier, these two requirements have also been incorporated into Article 4 of Section V.

Another major change to Section XI involved removing most of the detailed examination requirements from Article IWA-2000, expanding these requirements, and publishing them in Mandatory Appendices I, III, IV, and VI. Requirements for the initial calibration and periodic calibration checks on UT instruments were revised substantially to increase the intervals between calibration checks and permit the use of electronic calibration simulators. This change was made to eliminate total dependence on the large calibration blocks needed for many Section XI applications.

Projections and Trends for the 1990s

It is expected that the trend toward expanded criteria and requirements for qualifying and certifying NDE personnel will continue into the 1990s. Specifically, the philosophy of NDE system qualification via performance demonstration testing is expected to become more prevalent throughout the ASME Code rules. Currently, these requirements apply only to the UT personnel, equipment, and systems covered under ASME Section XI. However, similar requirements are currently being developed for Section XI ET applications, and such rules may also be considered for other Section XI NDE applications. Depending on the success of the Appendix VIII approach and the perceptions and reactions from industry and various regulatory bodies, the philosophy of performance demonstration could also be adopted by the other ASME Code sections.

Since the new Section XI requirements for multifrequency eddy current examination of steam generator tubes have now reached the final steps in the ASME Code approval process, these new Code rules are expected to become available in the early 1990s. Similarly, a new approach for conducting visual acuity testing will be incorporated into Section XI in the near future, and these may also be adopted by other Code sections.

A Code case on rules for using acoustic emission for inservice monitoring of nuclear components was published in 1989. This Code case provides explicit requirements for the personnel, equipment, procedures, and qualifications for monitoring predefined areas to detect crack growth during nuclear power plant operation. This Code case could ultimately lead to either a nonmandatory appendix in Section XI, a new article in Section V, or both. This would involve a broader scope of applicability and permit wider use of acoustic emission for monitoring the structural integrity of metallic pressure vessels during operation.

The philosophy of qualifying the NDE process as a system (i.e., the personnel, equipment, and procedures), rather than as individual elements, plus additional requirements for qualification via performance demonstration (using specimens or mockups that contain actual or simulated defects instead of calibration/reference blocks or other artificial test objects), is expected to see increasing use during the 1990s. This trend, initiated in Section XI for specific UT applications, will probably find increased usage for other Section XI needs and could also be adopted by Section III, Section V, and possibly even Section VIII for specific, critical applications. When and if the piping codes adopt this approach will depend on the perceived benefits realized from the nuclear and other applications.

Qualification of the RT process as a complete system (i.e., personnel, equipment, and procedures) via performance demonstration tests using real specimens with simulated or actual defects could ultimately be required for critical RT applications. Although radiography has traditionally utilized the IQI/penetrameter approach for qualifying and monitoring RT performance, adopting the philosophy of performance demonstration testing could also lead to improvements in the reliability of the radiographic method.

Technological advances in data compression and optical storage devices appear to have facilitated long-term storage of radiographic film images in digital format using optical media. This could offer a cost-effective alternative to meeting the current Code requirements for storage of radiographic records, while eliminating many of the problems inherent with long-term film storage. The major advantages of digital film storage include preservation of image quality, fast and efficient data retrieval, and lower storage costs. Limits on the use of gamma sources for section thicknesses below 1 in. and reducing the present limits on geometric unsharpness are changes that could also improve overall radiographic quality and may receive consideration during the 1990s.

It is expected that the UT technology will soon reach the point where reliable requirements for sizing and characterizing flaws can be developed and published in Section V and/or Section XI. Guidelines for computerized interpretation of UT indications should become available in the not too distant future, and these could lead to similar guidelines for interpreting indications produced by the other NDT methods. The use of sophisticated computerized UT imaging systems will probably increase dramatically during the 1990s.

With the availability of more accurate and reliable techniques for measuring contaminant levels in liquid penetrant materials, combined with the availability of higher purity raw chemicals, it is expected that consideration will be given to lowering the current 1% Code limits on the halogen and sulfur content in liquid penetrant materials.

Once the new rules covering the use of multifrequency eddy current equipment have been published in Section XI, it is expected that similar rules will be developed for Article 8 of Section V. This represents an urgent need since the current Section V rules for ET examination are technically obsolete by almost 10 years as we enter the decade of the 1990s.

In view of the importance of visual examination and the fact that ASNT is currently developing additional personnel qualification requirements for the VT method, it seems reasonable to expect that the Section V rules for visual examination (Article 9) will be expanded considerably beyond the current total of two pages.

As the various techniques become increasingly sophisticated and as the associated technologies become more advanced, it is expected that NDE data will become more quantifiable, hence more compatible with the concept of basing acceptance/rejection criteria on fracture mechanics theory and fitness-for-service criteria. This certainly has been a goal throughout the 1980s, and perhaps the technology will permit achieving that goal during the 1990s.

This goal is consistent with recent advances in the probabilistic risk assessment (PRA) approach for establishing inservice inspection criteria. Increased use of risk-based methods for developing ISI guidelines should produce ISI programs that maintain plant safety, while reducing the inspection costs. This approach should achieve widespread acceptance for applications where structural failures could cause significant human or economic loss such as nuclear and fossil-fueled power plants, petrochemical facilities, and large commercial aircraft.

Acknowledgments

The support and encouragement provided by the Battelle, Pacific Northwest Laboratory for my ASTM activities is gratefully acknowledged. A special word of thanks to Kay Hass for her conscientious and willing efforts during the preparation of this manuscript and the presentation upon which it is based.

References

[*1*] "Personnel of Codes, Standards, and Related Accreditation and Certification Committees," ASME AS-11, The American Society of Mechanical Engineers, New York, NY, January 1990.
[*2*] United States Code of Federal Regulations, 10CFR50 Part A.
[*3*] ASTM Charter, *ASTM 1990 Directory,* ASTM, Philadelphia, April 1990.
[*4*] Foreword to *1990 Annual Book of ASTM Standards,* American Society for Testing and Materials, Philadelphia.

Thomas D. Cooper[1] and Burl W. Nethercutt[2]

SAE/AMS NDT Standards

REFERENCE: Cooper, T. D. and Nethercutt, B. W., "SAE/AMS NDT Standards," *Nondestructive Testing Standards—Present and Future, ASTM STP 1151,* H. Berger and L. Mordfin, Eds., American Society for Testing and Materials, Philadelphia, 1992, pp. 153–162.

ABSTRACT: The development of nondestructive testing standards for the aerospace industry is a key task that is critical and that is receiving increasing attention. The Aerospace Materials Division of SAE International (formerly, the Society of Automotive Engineers) (SAE) has traditionally prepared and maintained a series of Aerospace Materials Specifications (AMS) covering the various NDT methods and materials. They were under the technical cognizance of Committee B, Finishes, Processes and Fluids. Because of the increasing importance of this area of technology, a new Committee (Committee K, Nondestructive Methods and Processes) was formed a few years ago to better focus attention on these documents and to ensure that the appropriate technical experts would be available to provide the current specifications that are needed. As the Department of Defense (DOD) has emphasized more reliance on nongovernment industry consensus standards bodies to begin to provide documents to replace military and federal specifications, this task has become even more urgent. Since its inception, Committee K has begun working on the task of upgrading older documents and creating new ones that are badly needed. This presentation will highlight the history and accomplishment of Committee K to date and will outline its future plans to provide AMS documents covering all of the major nondestructive inspection methods.

KEY WORDS: nondestructive testing, standards, specifications, nongovernment industry standards, SAE Committee K, Aerospace Materials Specifications (AMS)

To many people, SAE International (SAE) is synonymous with U.S. automobile technology. In fact, however, it is much more widely based than that and is devoted to developing, collecting, and disseminating mobility technology on a worldwide basis, as indicated in Fig. 1. SAE has a very strong activity in aerospace technology, and for over 50 years has prepared and published the widely known and respected Aerospace Materials Specifications (AMS), which are recognized and used on an international basis. It is the purpose of this paper to present a brief overview of the activities of SAE and in particular the Aerospace Materials Division of the Aerospace Council. Details will then be provided on the work of Committee K, Nondestructive Methods and Processes, and its role in the preparation of standards for worldwide use in nondestructive testing.

The structure of SAE is shown in Fig. 2. Of particular interest is the Technical Board, under which are located a number of councils and a technology development group which cover the broad spectrum of technical activities of the Society. Aerospace activities are centered under the Aerospace Council (Fig. 3), a 25-member body with international representation. The Council is currently chaired by C. Julian May, senior vice-president of Delta Airlines. The other 24 members of the Council are from the major aerospace companies of the United States

[1] Chief, Systems Support Division, Materials Directorate, Wright Laboratory, Wright-Patterson Air Force Base, OH 45433-6533, and chairman, AMS Division, SAE.

[2] Specialist, American Airlines, Maintenance and Engineering Center, P.O. Box 582809, MD 23, Tulsa, OK 74158, and Chairman, Committee K, AMS Division, SAE.

SAE INTERNATIONAL

Purpose:

To develop, collect and disseminate the knowledge of mobility technology on a worldwide basis in order to advance these fields and their practitioners in a manner which serves humanity.

FIG. 1—*SAE international purpose.*

and Europe, the major airlines and appropriate government agencies, such as the FAA, DOD, NASA, and the National Aeronautical Establishment of Canada. Industry representatives are generally at the engineering vice-presidential level or equivalent. The Council guides the activities of the seven major divisions that make up its structure as shown in Fig. 4.

The Aerospace Materials Division was created just over 50 years ago and has a long and proud history (Fig. 5). At that time, the Engine Technical Committee of the Aeronautical Chamber of Commerce of America (ACCA) formed its Materials Committee, composed of representatives of aircraft engine and propeller manufacturers. Shortly thereafter, the Committee recommended that its work be centralized in a recognized standardizing agency of the aircraft and associated materials industries. The ACCA proposed that the work be transferred

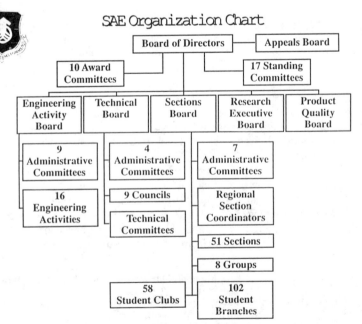

FIG. 2—*SAE organization chart.*

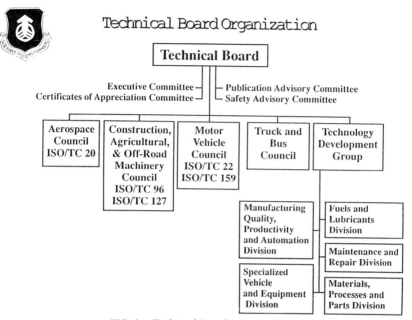

FIG. 3—*Technical Board organization.*

to the Society of Automotive Engineers, and, as a result, the SAE Council created the Aircraft Materials Division of the SAE Standards Committee in October 1939, using essentially the same membership as the ACCA Committee. The first Aeronautical Materials Specifications (AMS) were published in December 1939, covering a limited number of materials, processes, and parts used by the aircraft engine and propeller manufacturers. In the years since that time, activities have expanded to include airframe, space, accessories, and special equipment. In recognition of this expansion, the name was eventually changed to "Aerospace" instead of "Aeronautical" to reflect the significant broadening of scope.

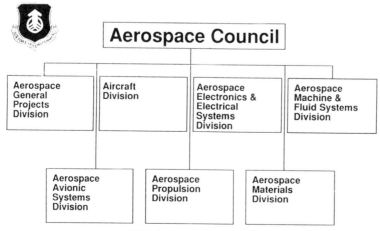

FIG. 4—*Aerospace Council.*

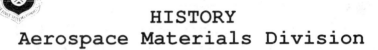

HISTORY
Aerospace Materials Division

- Created in 1939

- Started as the Materials Group, Engine Technology
  Committee, Aeronautical Chamber of Commerce
  of America

- Transferred to SAE as the Aircraft Materials Division

- First AMS published in December 1939

- Members: 8   ———————▶   >600

- Documents: 101   ———————▶   >2200

FIG. 5—*SAE Aerospace Materials Division/History.*

From a membership of eight when the original group was formed by the ACCA, the Division has grown to a membership in excess of 600. The original group of 101 documents has grown to over 2200 individual specifications which are constantly being reviewed and updated. In 1990, 300 new and revised specifications were issued.

An interesting and complete history of the evolution of Aerospace Materials Specifications and the Aerospace Materials Division of SAE was recently published by Hafeez [1–4], which covers in detail the creation and growth of this activity. The Division, shown in Fig. 6, currently consists of 13 commodity committees under three major groupings, which are responsible for the preparation and updating of the AMS documents. In addition to these publishing committees, there are several other activities, including the Aerospace Metals Engineering Committee, which is a technical advisory committee, the National Aerospace Defense Contractors Accreditation Program (NADCAP) Steering Committee, which is establishing a third-party accreditation program and several task forces covering specific technical topics. There are also a Coordinating Committee, which helps administer the Division, an Editorial Consultants Committee, which ensures consistency in the documents, and an Advisory Board. Of particular interest to this conference, however, is Committee K, Nondestructive Methods and Processes.

Committee K: Nondestructive Methods and Processes

Prior to 1982, the AMS documents dealing with aerospace nondestructive testing were prepared and updated by Committee B, the Finishes, Processes and Fluids Committee. Committee B was tasked with a wide variety of materials and processes and, lacking sufficient numbers of members and expertise to cover such a wide variety of topics, which include plating, coatings, NDT, joining, lubricants, corrosion prevention compounds, etc., the NDT documents were not widely accepted and used in the NDT aerospace community. At about that time (Fig. 7), a group of NDT specialists in the airline business was addressing the problem of generating a generic fluorescent penetrant processing specification that would be acceptable for use by all of the commercial airlines. This effort was being carried out by representatives of the airlines, airframe, and turbine engine manufacturers and was being worked in conjunction with the

SAE
Aerospace Materials Division

Division Advisory Board	Division Coordinating Committee
Military Task Force	Editorial Consultants Committee
NADCAP	Metric Advisory Panel

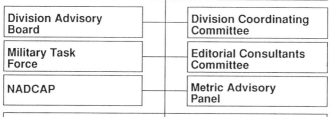

METALS	Non Metals and Chemical Processes	Nondestructive Inspection

B - Finishes, Processes, & Fluids
D - Non Ferrous Alloys
E - Carbon & Low Alloy Steels
F - Corrosion & Heat Resistant
  Alloys
G - Titanium, Beryllium, &
  Refractory Materials
L - Plating & Materials Coating
AMEC - Aerospace Metals
  Engineering
	CC - Composites
	CE - Elastomers
	CP - Polymers
	H - Electronic Materials &
	  Applications
	J - Aircraft Maintenance
	  Chemicals & Materials
	G-9 - Aerospace Sealing
		K - Nondestructive Methods
		  & Processes

FIG. 6—*SAE Aerospace Materials Division.*

Air Transport Association (ATA). Meetings of the group were generally held during the ATA's annual Nondestructive Testing Forum. As work on that document progressed, it became a natural connection for the group to be formalized under the SAE Aerospace Materials Division and to begin to address the AMS aerospace NDT documents across the board. Committee K was officially formed in March 1982, and since that time has assumed responsibility for all of the aerospace NDT documents formerly handled by Committee B.

COMMITTEE K
Nondestructive Inspection

- Ad Hoc Committee in Airline Industry (1982)

- Creating Generic Fluorescent Penetrant Processing Specification

- Airlines, Airframe and Engine Manufacturers

- Became Part of AMD in 1982

- Assumed Responsibility for All Aerospace NDT Documents

FIG. 7—*Committee K/Nondestructive Inspection.*

DOCUMENT GENERATION

- Documents Drafted or Updated by Technical Experts
- 28 Day Ballot Prepared by AMS Editorial Consultant
- Essential Comments Resolved
- 14 Day Ballot on Changes, If Necessary
- Submitted to Aerospace Council for Approval, Publication

FIG. 8—*Document generation.*

Committee K functions the same as the other commodity committees in the Division and prepares and issues AMS documents according to the Organization and Operations Guide of the Division [5]. The process (Fig. 8) basically consists of having a volunteer or volunteers serve as sponsor and either draft a new specification or update an existing specification, which must be done at least once each five years (ANSI requirement). When the document is drafted, it is sent to the appropriate AMS editorial consultant, who prepares it in the form of a 28-day ballot. The ballot is circulated to all members and supplier consultants for their review and comment. The results are returned to the sponsor for review and resolution of any essential comments. Those essential comments that cannot be resolved are placed on the agenda for discussion at the next meeting of the Committee. At that time, the issues are resolved by voice vote and a 14-day ballot is sent out documenting those technical changes that were agreed upon. Only technical changes made to the original document are subject to comment. If no issues are raised on the 14-day ballot, the document is prepared by the editorial consultant for submission to the Aerospace Council for final approval and publication. If further issues are raised, they will be resolved at the next meeting and another 14-day ballot issued.

According to the charter of the Division (Fig. 9), only aerospace users are qualified to be voting members, insuring that the Division remains as a user-controlled activity. However, because all essential technical comments are welcome and must be addressed, anyone attending the meeting is encouraged and allowed to vote. If a particularly critical issue is raised, however, the chairman can call for a "users only" vote, which occasionally occurs. This total process assures that the resulting documents are "user controlled," but at the same time allows all interested parties the opportunity to participate and also insures that "due process" is afforded to everyone.

Membership on Committee K currently consists of 32 voting members, supplemented with 7 consultant members, 13 consultant supplier members, and an additional 43 mailing-list-only individuals. Meetings are held twice a year, in the spring concurrently with the ASNT spring conference and in the fall concurrently with the ATA NDT forum. Meetings of special task forces under the Committee are held more frequently.

At the present time, there are 22 aerospace materials specifications (AMSs), two Aerospace recommended practices (ARPs), and one aerospace standard (AS) for which Committee K is responsible. They are listed in Table 1. In addition to the documents listed, there are two active task forces that are working on additional documents. They are the Magnetic Particle/Liquid Penetrant Task Force, which has several documents in preparation, and the Inspection of Ground Chrome Plated Surfaces Task Force, which is preparing a single document.

CURRENT ACTIVITIES

- User Controlled Activity

- Current Membership
  - 32 Voting Members
  - 7 Consultant Members
  - 13 Consultant Supplier Members

- Two Meetings Per Year

- Workload
  - 22 Aerospace Materials Specifications (AMS)
  - 2 Aerospace Recommended Practices (ARP)
  - 1 Aerospace Standard (AS)

- Action Task Forces
  - Magnetic Particle/Liquid Penetrant
  - Inspection of Ground Chrome Plated Surfaces

FIG. 9—*Current activities.*

The task that faced Committee K when it was created was formidable indeed. The number of documents badly in need of revision was long, and the need for additional documents was urgent. The first major new specification generated was AMS-2647, Fluorescent Penetrant Inspection, Aircraft and Engine Component Maintenance. This specification allowed the major aircraft manufacturers to have a common penetrant inspection process which, in turn, has been provided to the airlines for their use. This has resulted in uniformity throughout the industry, allowing the airlines to have a single penetrant process specification for application to all of their equipment. This has been a major step forward. Currently, Committee K has eight AMS documents under review.

There are other factors at work now (Fig. 10) that impact the future activities of Committee K. For a number of years, it has been the policy of the Office of Management and Budget

OTHER FACTORS

- Office of Management and Budget (OMB) Policy

- Support Non-Government Standard Bodies (NGSB) Activities

- Replace Military and Federal Documents When Possible

- Committee K Currently Targeting Appropriate NDT Documents

  - MIL-I-25135

FIG. 10—*Other factors.*

TABLE 1—*Society of Automotive Engineers (SAE) Aerospace Materials Division Committee K, Nondestructive Methods and Processes Documents.*

		AEROSPACE MATERIALS SPECIFICATIONS	
AMS	2630A	UT Inspection, Products over 0.5 in. thick	(Apr 80)
	2631A	UT Inspection, Ti and Ti Alloy Bar and Billet	(Jan 85)
	2632	UT Inspection of Thin Materials, 0.5 in. or less	(Mar 74)
	2633A	UT Inspection Centrifugally-Cast, Corrosion Resistant Steel Tubular Cylinders	(Oct 89)
	2634	UT Inspection, Thin Wall Metal Tubing	(Apr 80)
	2635C	Radiographic Inspection	(Jul 81)
	2640J	Magnetic Particle Inspection	(Oct 83)
	2641	Vehicle, Magnetic Particle Inspection, Petroleum Base	(Jan 88)
	2645H	Fluorescent Penetrant Inspection	(Jan 83)
	2646C	Contrast Dye Penetrant Inspection	(Apr 82)
	2647	Fluorescent Penetrant Inspection, Aircraft Engine Component Maintenance	(Apr 85)
	3040A	Magnetic Particles, Nonfluorescent, Dry Method	(Jul 88)
	3041B	Magnetic Particles, Nonfluorescent, Wet Method, Oil-Vehicle, Ready to Use	(Jul 88)
	3042B	Magnetic Particles, Nonfluorescent, Wet Method, Dry Powder	(Jul 88)
	3043A	Magnetic Particles, Nonfluorescent, Wet Method, Oil Vehicle, Aerosol Packaged	(Jul 88)
	3044C	Magnetic Particles, Fluorescent, Wet Method, Dry Powder	(Jul 89)
	3045B	Magnetic Particles, Fluorescent, Wet Method, Oil Vehicle, Ready-to-Use	(Jul 89)
	3046B	Magnetic Particles, Fluorescent, Wet Method, Oil Vehicle, Aerosol Package	(Jul 89)
	3155C	Oil, Fluorescent Penetrant, Solvent-Soluble	(Jul 83)
	3156C	Oil, Fluorescent Penetrant, Water Washable	(Oct 83)
	3157B	Oil, Fluorescent Penetrant, High Fluorescence, Solvent Soluble	(Oct 80)
	3158A	Solution, Fluorescent Penetrant, Water Base	(Jul 79)
		AEROSPACE RECOMMENDED PRACTICES	
ARP	1333	Nondestructive Testing of Electron Beam Welded Joints in Tri-Base Alloys	(Mar 74)
	4462	Barkhausen Noise Inspection for Detecting Grinding Burn in High Strength Steel Parts	(Jan 91)
		AEROSPACE STANDARDS	
AS	1613A	Image Quality Indicator, Radiographics	(Jan 88)

(OMB) that the government should, to the maximum extent possible, get out of the specifications and standards preparation business. The DOD has actively supported this policy by strongly encouraging nongovernment standards bodies (NGSBs) to prepare industry consensus documents that can eventually replace military and federal specifications and standards. This policy is clearly stated in a recent report to the Secretary of Defense by the Under Secretary of Defense (Acquisition) [6]. Part of that policy includes encouraging technical personnel of the DOD to play an active role with NGSBs to insure that DOD's needs are adequately represented in the resulting documents. Consequently, one of the future goals of Committee

CONCLUSIONS

- SAE Has 50 Year History of Activity in Aerospace Specifications and Standards Worldwide

- Focused Activity in NDT Since 1982 Under Committee K

- Activities Under Way

    •• Prepare Documents to Replace MIL and FED Documents

    •• Generate High Quality Documents for National and International Use

FIG. 11—*Conclusions.*

K is to target those critical military and federal specifications and standards concerning NDT materials and processes and prepare replacement documents that will be acceptable to DOD for future use. The first major specification undertaken is MIL-I-25135. This specification is of great importance because it covers the materials used in liquid penetrant inspection of metal and nonmetal components. This specification is critical to penetrant manufacturers because it sets the standards with which their products must comply. It is recognized that the transition from government to industry consensus specifications and standards will not occur overnight, but will be a long-term on-going activity.

Conclusions

SAE has been active for over 50 years in the development of both national and international standards, many of which cover aerospace nondestructive testing methods and materials (Fig. 11). Since 1982, with the formation of Committee K, renewed emphasis has been placed on this technology area, with the result that a number of new and revised documents have been issued. With the increased pressure to replace military and federal specifications and standards by industry consensus documents developed by appropriate nongovernment standards bodies (NGSB), the challenge has been presented to SAE and Committee K to get on with the process of generating high-quality documents that will be accepted and used by not only the U.S. aerospace industry, but on an international basis as well.

References

[1] Hafeez, A. and Shopp, W., "SAE and AMS—A Half Century of Aerospace Excellence," Part I: "History of AMS," *Aerospace Engineering*, June 1989, pp. 23–25.
[2] Hafeez, A. and Shopp, W., "SAE and AMS—A Half Century of Aerospace Excellence," Part II: "AMS Development and the Aerospace Materials Division," *Aerospace Engineering*, July 1989, pp. 21–23.
[3] Hafeez, A. and Shopp, W., "SAE and AMS—A Half Century of Aerospace Excellence," Part III: "AMS Documents," *Aerospace Engineering*, August 1989, pp. 15–17.
[4] Hafeez, A., "SAE and AMS—A Half Century of Aerospace Excellence," Part IV: "The Future of SAE and AMS," *Aerospace Engineering,* September 1989, pp. 17–20.

[5] "The SAE Aerospace Materials Division Organization and Operation Guide," July 1984, SAE, Warrendale, PA.

[6] "Enhancing Defense Standardization—Specifications and Standards: Cornerstones of Quality," report to the Secretary of Defense by the Under Secretary of Defense (Acquisition), November 1988, Office of the Assistant Secretary of Defense, Defense Quality and Standardization Office, Falls Church, Virginia 22041–3466.

Leonard Mordfin[1]

U.S. Participation in the Development of International Standards for Nondestructive Testing

REFERENCE: Mordfin, L., "**U.S. Participation in the Development of International Standards for Nondestructive Testing,**" *Nondestructive Testing Standards—Present and Future, ASTM STP 1151,* H. Berger and L. Mordfin, Eds., American Society for Testing and Materials, Philadelphia, 1992, pp. 163–168.

ABSTRACT: It is important to the U.S. balance of trade that international standards for nondestructive testing (NDT) be compatible with the practices used in this country. This need is becoming more critical as the European Community approaches its goal of a single market with uniform standards. Unfortunately, the private sector of the U.S. economy does not appear to have adequately recognized this need, nor has the public sector faced up to it, so that U.S. participation in the development of international standards for NDT has been less than adequate. The nature and the status of this participation are explained and discussed.

KEY WORDS: ISO, nondestructive testing, standards

In 1977 the United States voluntarily relinquished the position it had held as the secretariat of the technical committee on NDT in the International Organization for Standardization (ISO). Up to that point, this committee (TC 135) had promulgated only three standards, none of which were concerned with the principal NDT methods, and none of which contained any measurable U.S. input. Thus, although the country that administers an ISO technical committee is in a position to exert a strong influence on the output of the committee, the U.S. withdrawal from this position did not arouse any significant protest from the domestic NDT community. On the contrary, it might be said that there was a general feeling of smugness in this community; a belief that this country's NDT standards and its NDT personnel certification system are the best in the world, so that there is little to be gained from participation in the development of international standards.

Unfortunately, many of our manufacturing-related standards—including those for NDT— became suspect when the inadequate quality of some American products began to be exposed in the late 1970s and early 1980s. At that point, one would imagine, U.S. industry would immediately have initiated an intensive effort to collaborate with the rest of the world in the development of international standards for NDT, to get U.S. input into those documents, so that the requirements by other countries for NDT on the products they buy would be consistent with the practices used in this country. Incredibly, such has not been the case. For the most part, either through lack of awareness or shortsightedness, American manufacturers continue to forego opportunities to influence the development of international standards for NDT.

[1] Group leader, Mechanical Properties and Performance, National Institute of Standards and Technology, Bldg. 223/Room B144, Gaithersburg, MD 20899.

The situation has now become critical. The common European internal market will become a reality by the end of 1992, and products which do not conform to the European Community's standards will likely be excluded from that market. The American National Standards Institute (ANSI), the U.S. member body in ISO, has negotiated agreements with the European standardization committees to permit U.S. input into the Community's standards, but this path will be narrow and indirect at best. On the other hand, the Community is committed to accepting ISO standards and would even forego the development of new European standards if a comparable and significant ISO activity were already in process.

Until recently, there was some token support available from the Federal Government to bring this message to the industrial community and to help coordinate the little participation that could be mustered from the private sector. Now, even that is drying up.

For machinery, structures, and other products for which preservice NDT is vital, the future for American industry in European markets would be bleak, indeed, were it not for a small band of conscientious and concerned experts who have taken it upon themselves, with only the most inadequate support, to help fashion proposals for international NDT standards based on the best practices of American industry. This cadre of dedicated individuals serves as the core of the U.S. Technical Advisory Group (TAG) to ANSI for ISO Technical Committee 135 on Non-Destructive Testing. Within the past few years alone, no fewer than eleven full-length American draft standards in ISO format have been launched on the long and tortuous path through the ISO system for, hopefully, eventual adoption as international standards.

This paper documents the status of this determined effort, against huge odds, to provide some measure of U.S. input into international standards for NDT. ASTM standards mentioned in this article are shown in Table 1.

Organization

Before proceeding with the status reports, it is helpful to clarify the roles that ANSI, ASTM, and NIST play in the development of international standards for NDT. On the basis of many inquiries that are received, it is clear that these roles are not well understood by most of those who have had no direct involvement in these activities.

The members of ISO TC 135 on NDT are countries; ANSI is the U.S. member body. When responding to letter ballots, or when its delegates participate in meetings of the Committee, ANSI is responsible for voting in the best interests of the United States. In order to do this, ANSI authorized the establishment of the TAG on NDT. The principal responsibility of the TAG is to determine the consensus U.S. position on matters pertaining to NDT which are being discussed and balloted in TC 135 and its subcommittees, and to advise ANSI and its delegates so that the U.S. position may be represented accordingly.

The TAG consists of its members and the TAG administrator. ANSI designated ASTM to serve as the TAG administrator so that, in effect, ANSI's direct role in regard to TC 135 is simply as a conduit for communications. Ballots, meeting announcements and agendas, meeting minutes, etc. are received by ANSI from ISO and from other member bodies of ISO and are relayed to the TAG. The TAG, in turn, determines how the U.S. vote should be cast on the ballots and who should represent the U.S. at meetings, and provides this information to ANSI for implementation.

TC 135 currently has six active subcommittees. Each subcommittee, as well as TC 135 itself, has a secretariat, that is, a country designated to administer the business affairs of the subcommittee or committee. Thus, the U.S. is the secretariat for Subcommittee 3 on Acoustical Methods. Here, too, ANSI has delegated the responsibility for providing the secretariat function for this subcommittee to ASTM. Lest there be any underestimation of ANSI's role in interna-

TABLE 1—*List of ASTM standards.*

Designation	Title
E 94	Guide for Radiographic Testing
E 165	Practice for Liquid Penetrant Inspection Method
E 545	Method for Determining Image Quality in Direct Thermal Neutron Radiographic Testing
E 569	Recommended Practice for Acoustic Emission Monitoring of Structures During Controlled Stimulation
E 610	Terminology Relating to Acoustic Emission
E 746	Method for Determining Relative Image Quality Response of Industrial Radiographic Film
E 748	Practices for Thermal Neutron Radiography of Materials
E 750	Practice for Measuring Operating Characteristics of Acoustic Emission Instrumentation
E 803	Method for Determining the L/D Ratio of Neutron Radiography Beams
E 1001	Practice for Detection and Evaluation of Discontinuities by the Immersed Pulse-Echo Ultrasonic Method Using Longitudinal Waves
E 1025	Practice for Hole Type Image Quality Indicators Used for Radiography
E 1030	Test Method for Radiographic Testings of Metallic Castings
E 1032	Method for Radiographic Examination of Weldments
E 1106	Test Method for Primary Calibration of Acoustic Emission Sensors
E 1254	Guide for the Storage of Radiographs and Unexposed Industrial Radiographic Films
E 1324	Guide for Measuring Some Electronic Characteristics of Ultrasonic Examination Instruments

tional standardization, it is appropriate to point out here that ANSI represents the United States in the highest councils and policy-making bodies of ISO and has been particularly effective in negotiating American access to the European Community's standardization system.

The National Institute of Standards and Technology (NIST) has no direct role in ISO's standardization activities other than voluntarily providing technical expertise, through its staff, as it does for domestic standardizing bodies. That the chairmen of the TAG and of Subcommittee 3 are both NIST employees is largely coincidental. These positions are not reserved for NIST staff; previous chairmen have been from industry.

Status Reports

Surface Methods

In 1984, two international standards on the liquid penetrant method were promulgated through the efforts of TC 135: ISO 3452 on the general principles of penetrant inspection, and ISO 3453 on the means of verifying liquid penetrant inspections. Neither of these documents had been introduced by the United States, but U.S. input and comments by consultant C. W. McKee and others throughout the standards development process assured that they are technically equivalent, at least in part, to ASTM E 165, and that they are not seriously inconsistent with good American practices. McKee now represents the United States on an international working group which was established to update the two standards in preparation for their reconfirmation.

Three other documents on surface methods are under development in TC 135: one on penetrant flaw detectors, one on magnetic particle testing equipment, and one on the magnetic particle test method. U.S. experts are participating in this activity to assure that the resulting standards will be acceptable to U.S. industry.

Ultrasonics

TC 135 has not yet promulgated any international standards dealing with ultrasonic NDT, but one document has reached the committee draft stage (formerly called draft proposal). This is a test method for the characterization of ultrasonic search units and sound fields, based largely on ASTM standards, which was prepared by M. C. Tsao of Ultra Image International, SAIC. Two other working drafts on ultrasonic NDT have also been introduced into the ISO system by the United States. These are technically equivalent to ASTM E 1001 on immersion testing and ASTM E 1324 on the characterization of ultrasonic instrumentation. J. D. Fenton, formerly of LTV Aerospace, is credited with recasting these ASTM standards into ISO format. (This is not a trivial exercise. Converting an ASTM or other standard into the ISO format can be quite tedious, but it can also be rewarding. It is not difficult to understand the reluctance— perhaps subconscious—of a delegate from one country to accept a standard from another country, particularly if there are political or economic differences between the two countries, even though such matters are not supposed to enter into ISO activities. Sanitizing a national standard by rewriting it in the ISO format before submitting it for international consideration is good practice.)

Tsao is presently drafting a new ISO working document on ultrasonic reference blocks.

Acoustic Emission

Three working drafts on acoustic emission topics, submitted by the United States, are under consideration by TC 135. These are based, respectively, on ASTM Standards E 610 on terminology, E 750 on the characterization of acoustic emission instrumentation, and E 1106 on the primary calibration of acoustic emission sensors. Here, too, the conversion of the ASTM documents into the ISO format is attributed to Fenton. Still another working draft, this one based on E 569 on the monitoring of structures during controlled stimulation, is in preparation.

Within TC 135, work on ultrasonic NDT and on acoustic emission NDT are both carried out in Subcommittee 3 on Acoustical Methods. This is the only one of the Committee's six current subcommittees for which the United States holds the secretariat; D. G. Eitzen of NIST is the chairman and G. A. Luciw of ASTM is the secretary. In principle, this might be seen as an opportunity for the United States to accelerate its documents through the system, particularly since ISO procedures are not nearly as rigorous as ASTM's with regard to the handling of negative votes and comments. In practice, the opposite has been the case. Ever concerned with avoiding even an appearance of partiality, Tsao, Eitzen, and Luciw repeatedly invite comments on U.S.-generated drafts in Subcommittee 3 and meticulously address them all before attempting to ballot a document for advancement to the next higher level of ISO consideration.

X-Ray Radiography

ISO presently has two standards on X-ray radiography that were developed by TC 135, both promulgated in 1985. ISO 5579 on the basic rules for radiographic examination of metallic materials is related, but not technically equivalent, to parts of ASTM Standards E 94, E 1030, and E 1032. John Munro III of RTS Technology serves as the U.S. expert on a working group charged with revising this standard in preparation for its reconfirmation. ISO 5580 on the minimum requirements for radiographic illuminators is an interesting example of the flow of standards technology in a reverse direction . . . from ISO to ASTM. Although the United States had only minor input into the original formulation of this international standard, its usefulness was subsequently recognized here, and an ASTM standard based on ISO 5580 is now under development.

Three ASTM standards on radiographic topics were converted into the ISO format—again through the diligent and untiring efforts of J. D. Fenton with help from H. C. Graber of Babcock & Wilcox—and have been introduced into the ISO system. These are E 746 on determining the relative image quality response of radiographic film, E 1025 on hole-type image quality indicators, and E 1254 on storage of radiographs and films.

The development of radiographic standards in TC 135 is carried out in Subcommittee 5 on Radiation Methods. The chairman of this subcommittee, H. Heidt of Germany, has indicated that he views the ISO version of ASTM E 746 as one part of an eventual two-part international standard on the classification of X-ray film, the other part being a document submitted by Germany.

Neutron Radiography

The development of international standards for NDT by neutron radiography is carried out in TC 135 by a working group of Subcommittee 5 on Radiation Methods. The U.S. representative on this working group, J. S. Brenizer of the University of Virginia, was recently appointed its convenor (i.e., chairman). The group has been very industrious, meeting once or twice a year since its establishment in 1987. Three ASTM standards are under consideration by the group: E 747 on standard practices for thermal neutron radiography of materials, E 545 on determining the L/D ratio of neutron radiography beams. The first of these, E 748, was submitted in ISO format and appears to be nearing acceptance as an ISO committee draft.

Other Methods

Two subcommittees of TC 135 are comparatively inactive at this time. These are Subcommittee 4 on Eddy Current Methods and Subcommittee 6 on Leak Detection Methods. Following a meeting of these two subcommittees in 1988, and in response to requests from the chairmen of the subcommittees, the United States submitted proposals for new work items, but there has been no response. This situation is potentially detrimental to U.S. interests; the opportunity to promote the development of international standards in these two important areas of NDT prior to the European Community's economic unification is rapidly slipping away. The U.S. delegation plans to raise this subject at the next meeting of TC 135, scheduled for 14–17 May 1991.

Personnel Qualification

The effort to establish an ISO standard for the qualification and certification of NDT personnel has generated considerable interest over the past several years. A draft international standard on this subject, ISO DIS 9712, which the United States vigorously opposed, was narrowly defeated in a recent international ballot. The Canadian secretariat of Subcommittee 7, which has jurisdiction over this standardization activity in TC 135, announced its intention to revise the document in order to overcome some of the objections. The United States, however, is seeking to introduce its own proposal, namely, the new ASNT standard on this subject, for consideration in lieu of DIS 9712.

Concluding Remarks

An attempt has been made in this paper to convey the importance to the United States of pursuing the development of international standards for NDT that are consistent with the best practices of American industry. Despite this importance, support for this effort is both meager and diminishing. Nevertheless, the progress which has been made by the U.S. TAG over the past few years has been significant, primarily as a result of individual and voluntary efforts by a handful of concerned and enlightened experts; but so much more could be done and so little time remains. Hopefully, this paper will serve to convey this important message to a broader audience than heretofore and yield some much needed new support for the TAG, either financial or in the time and efforts of volunteers. Inquiries are always welcome and may be directed to the author or to ASTM.

Acknowledgment

The point has been made, perhaps excessively, that support for this work from either governmental or industrial sources has been very inadequate considering both the significance and the magnitude of the job to be done. It is essential, therefore, to point out a notable exception, and that is the role of ASTM. With nothing to gain from the promulgation of ISO standards and, indeed, a possibility of significant loss, ASTM has actively supported what it perceives to be the best interests of the United States in regard to the work of TC 135. Largely through the enthusiastic and competent efforts of staff manager George Luciw, ASTM has administered the TAG and provided the secretariat function for Subcommittee 3 at considerable expense, to say nothing of covering the substantial fees imposed by ANSI (and by ISO through ANSI) for these activities. For this, the members of the TAG are sincerely grateful to the officers and directors of ASTM. Without ASTM's contribution (and support from ASNT for the work on personnel qualification) there would be no U.S. participation in the development of international standards for NDT. It is hoped that the country's NDT community will one day recognize the importance of this contribution.

Elie E. Borloo[1]

Harmonization of NDT Standards Within the European Economic Community

REFERENCE: Borloo, E. E., "Harmonization of NDT Standards Within the European Economic Community," *Nondestructive Testing Standards—Present and Future, ASTM STP 1151,* H. Berger and L. Mordfin, Eds., American Society for Testing and Materials, Philadelphia, 1992, pp. 169–179.

ABSTRACT: The European Council of Ministers invited the Commission of the European Communities (CEC) to draw up a policy on conformity assessment involving testing and calibration, quality assurance, certification, and accreditation, aiming to *harmonize* the existing national regulations in order to eliminate barriers to trade, principally after 1 Jan. 1993 (beginning of the Single Market).

The CEC gave mandates to European standardization bodies to draw up European standards for Community purposes.

To date, several standardization programs have been mandated under this system; in particular, one of the ongoing works at CEN (Comité Européen de Normalisation) level is in the field of nondestructive testing—it covers the pressure vessels area.

1. CEN TC 138 is the technical committee (TC) involved with nondestructive testing. This TC has the scope to set up standards on terminology, equipment, and general principles of methods for the different NDT methods as well as to standardize the principles of qualification and certification of NDE testing personnel.
2. CEN TC 121 is the TC involved with welding. This TC has a working group, 5B, involved with NDT and six subworking groups, the scope of which is to set up standards on general rules, examination techniques, and acceptance criteria for the different NDT techniques when examining welded structures.

Participants in the different working groups are experts from the 12 EEC countries and also from the six EFTA countries. These experts are technicians from industry, technical laboratories, or from national standardization bodies.

1. EEC: European Economic Community: Belgium-Denmark-France-Germany-Greece-Ireland-Italy-Luxembourg-Portugal-Spain-The Netherlands-United Kingdom.
2. EFTA: European Free Trade Association: Austria-Finland-Iceland-Norway-Sweden-Switzerland.

During these technical meetings, national and international, existing documents (e.g., DIN-AFNOR-BS-ASTM-ISO-IIW) are considered, aiming to create future European NDE standards.

KEY WORDS: European NDT standards, harmonization of national regulations, Single Market, prenormalization, certification

Among the fundamental rules of the EC Treaty is the principle of the free movement of goods inside the Community established by Articles 30 to 36. These articles prohibit member states from applying quantitative restrictions on imports unless such restrictions are justified

---

[1] Engineer of the Commission of the European Communities, Joint Research Centre, Ispra, Varese, Italy. Institute for Advanced Materials, Non-Destructive Evaluation of Materials Service.

on grounds foreseen by Article 36, such as public security or the protection of health and safety.

Where national regulations, acceptable on the grounds of Article 36, create barriers to intra-community trade, they can be eliminated by the harmonization of those national provisions. The instrument used by the Commission for Technical Harmonization is a directive of the Council. It is a document, proposed by the Commission and adopted by the EC Council of Ministers, containing technical requirements which ensure an appropriate level of safety for the products covered. Products complying with the directive thereby gain the right of free circulation in the Community.

To accelerate the harmonization work, the Council Resolution of 7 May 1985 was adopted describing a new approach to technical harmonization. Under this new approach, directives shall only establish the essential safety requirements goods must satisfy and make general reference to European standards for technical specifications. Furthermore, the Single European Act, adopted in 1986, contains the commitment of the European Economic Community to create, before the end of 1992, a single market, i.e., "an area without internal frontiers in which free movement of goods, persons, services and capital is ensured."

For many years and notwithstanding Article 30 of the Treaty, the free circulation of goods in the European Community has been hindered by the existence of differing national regulations, justified by the need to protect health and safety. Since the adoption of the Council Resolution of 7 May 1985, the harmonization work has been based on the principles of the "new approach." Directives, therefore, only specify essential safety requirements which products must satisfy, formulated in terms of performance, and European standardization bodies are given the task of establishing technical specifications, i.e., standards, which satisfy these requirements. The application of these to a product gives that product the benefit of the presumption of conformity to the mandatory essential requirement, but these standards remain optional. The mandatory directives specify that the manufacturer has the choice of not manufacturing in conformity with the standards but that in this event he has to prove that his products nevertheless conform to the essential requirements. A major advantage of this approach is that the mandatory essential requirements are formulated in general performance terms, with the result that technical innovations are not impeded and technical specifications, established by standards, can easily be adapted to take account of technical progress.

Standardization in Support of Directives

New approach directives need to be complemented by standards for two main reasons: first, the safety requirements in the new approach directives are formulated in quite general terms and need to be specified more to ensure a uniform interpretation. The corresponding standards, therefore, represent a "reference" for manufacturers as well as for certification bodies when they have to evaluate the conformity of the manufactured products with the safety requirements; second, the application of standards gives a presumption of conformity with the essential requirements. This means that manufacturers applying the standards do not have to demonstrate the compliance of the manufactured products with the requirements but only have to indicate that the technical standards which correspond to the essential requirements have been correctly applied.

The Commission has asked that a large amount of standardization work be carried out by European standardization bodies CEN/CENELEC/ETSI. This body will establish, on the basis of the mandate, a work program and fix the timetable by which standards shall be adopted. It should be noted that the EFTA countries, whose national standardization bodies are also members of the European standardization organizations, have consistently supported the Community's approach and contribute to the financing of mandated standardization work.

European Standardization Organizations

National Standardization Structures

National standards institutions in European countries were mostly set up between the two world wars. There are great similarities in their statutes and operating methods: they are almost always organizations set up by trade associations, and the documents elaborated by them (standards) are recognized by national authorities.

Unlike the situation in the United States and Canada, where several hundred organizations publish sectorial standards, in European countries the structures are centralized. Differences exist from one European country to another in:

1. Financial resources (contribution from industry, role of standards, public subsidies).
2. Extent of responsibility in writing standards (within the association as in France or at the national level as in the UK).
3. Dependence on national public authorities (fully independent as in Switzerland, part of Ministry as in Portugal).
4. Size importance of the institutions.

However, the similarities in European countries were such that European standardization bodies such as CEN (European Committee for Standardization), CENELEC (European Committee for Electrotechnical Standardization), and ETSI (European Telecommunications Standardization Institute) could be proposed.

CEN and Its Work Procedures

The European Committee for Standardization (CEN) is a nonprofitmaking and international association of the national standards organizations of twelve countries of the European Community (EC) and of the six countries of the European Free Trade Association (EFTA), which are also members of ISO, the International Standardization Organization. The principal task of CEN is to prepare European standards (EN) that comprise a set of technical specifications established in collaboration with and with the approval of all parties concerned in the various member countries. The standards are established on the principle of concensus and adopted by the votes of a weighted majority. Adopted standards must be implemented in their entirety as national standards by all members regardless of the way in which they voted, and any conflicting national standard must be withdrawn.

CEN and CENELEC have in recent years recognized the value of using the services of other organizations, the so-called Associated Standardization Bodies (ASBs), in the preparation of technical documents destined to become European standards. A number of such bodies have been given this status, such as ECISS (European Committee on Iron and Steel Standardization) and AECMA (Association Européenne des Constructeurs de Matérial Aérospatial) and have been responsible for the programming and drafting of documents which have only to be submitted to inquiry and voting by CEN or CENELEC before becoming European standards. Approximately 100 European standards so far adopted by CEN and CENELEC have been provided by ASBs. The CEN organization is outlined in Fig. 1.

The General Assembly is composed of representatives from all member states; it is supported by an administrative board and a central secretariat. The preparation of new CEN standards as well as the revision of existing standards is carried out by a number of technical committees (TC).

Many new standards are needed to provide the technical basis for the free circulation of goods. The Technical Board (TB), therefore, has two main sources for new work items:

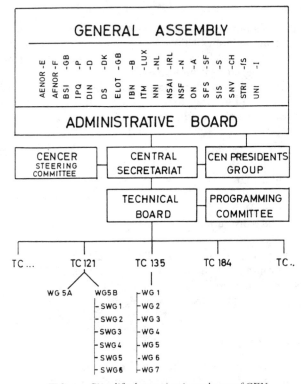

FIG. 1—*Simplified organization scheme of CEN.*

1. Ideas originating in the TC.
2. Standards ordered through a mandate of the CEC.

The close link between CEN standards and EEC requests necessitates an efficient coordination. Preparation of a new standard or a new work item has to be carried out on the basis of a detailed scope of work. An important part of the scope of work is a reference to the document(s) [often international or national standard(s)] on which the work shall be based. The scope of work must be approved by the Technical Board (see Fig. 1) before the technical committee can start actual work.

To ensure consistent planning, programming, and coordination of European standardization activities within a particular sector, CEN/CENELEC may also set up programming committees responsible for drawing up a European standardization program. This is an innovation introduced by the new internal regulations so as to ensure that the priorities for the unification of Europe are taken into account. The members of these committees are as far as possible chosen from circles representative of the main involved interests.

Once a new work item has been approved, all activities are controlled directly by the technical board within the limits set by the approved scopes of work and the corresponding time schedules. The Central Secretariat is kept informed on the progress of the work.

A technical committee may establish one or more working groups. A working group usually has the task of preparing a draft for one or more standards (in accordance with an approved scope of work), and the group is disbanded once these standards have been accepted as European standards. The working group may hold frequent and long meetings; there are no restrictions. An efficient working group supported by a productive secretariat (from a national standardization organization) has a good chance of obtaining results in the interval between two plenary meetings of the parent technical committee. Therefore, CEN should be able to produce standards at a reasonable speed.

When a working group has prepared a draft standard, the document is presented to the parent technical committee for approval. The draft then is transmitted to the Technical Board for confirmation. The document is translated so that identical versions are available in the three official languages of CEN: English, German, and French.

After confirmation by the Technical Board, the standard is circulated to all CEN member countries for inquiry and voting during a six-month period.

The inquiry may result in technical remarks, which may necessitate further work. When the final version is available, it is sent to all CEN member states for formal voting during a two-month period. This is just to ratify the final version; technical remarks are not expected. This procedure may be shortened for preexisting documents, such as ISO standards. Essentially such documents are circulated for formal voting only.

CEN has a rather peculiar method for publication of the final standards. All CEN standards are published and implemented as national standards. Each national standardization organization has to perform a translation (if needed) of the CEN standard into the national language and publish and implement the CEN standard as a national standard. This has to be done not later than six months after approval (by formal voting) of the final CEN standard; at the same time, the national standardization organization has to withdraw any national standard not compatible with the CEN standard.

Application of CEN standards will be mandatory for industrial sectors regulated by directives from the EEC. Once a directive has been approved, national laws have to be modified accordingly.

CEN rules include a standstill agreement, which essentially means that national standardization organizations have to stop preparation of new or revised standards in areas where a CEN committee is working.

Other documents established by CEN are harmonized documents (HD) and European prestandards (ENV). A harmonized document is drawn up and adopted in the same way as a European standard, but its application is more flexible so that specific circumstances pertaining to some countries can be taken into account. No conflicting national standards may continue to exist after the date fixed by CEN. A European prestandard is prepared using a simplified procedure as a prospective standard for provisional application in technical areas where there is a high level of innovation or an urgent need for guidance. CEN members are required to make the ENV available at the national level, but differing national standards may still be kept in parallel.

The harmonization work is monitored by a Technical Board which follows the recommendations made by competent programming bodies. The work, based as far as possible on the results of international standardization or where appropriate on other sources, is performed by technical committees (TC) composed of technical experts by the national standardization organizations.

Some working documents may, prior to reaching the prEN stage, follow the so-called Questionnaire Procedure (QP). This procedure permits the Technical Board to find out the existing degree of national harmonization with the reference document in question and to assess the acceptability of that document as EN or HD. Approval of an EN, HD, or ENV results from a vote in which the CEN/CENELEC members are all entitled to take part. The voting rules require that any of these publications are based on sufficient agreement among the represented countries in order to ensure the widest possible applicability of its content. In addition to the requirement for a simple majority, the votes shall be accorded specific weightings. Each country is allocated a number of votes as specified in Table 1. Minimum conditions for the approval of a draft standard as CEN standard by weighted voting are the following. Votes from all members are counted first, and the proposal shall be adopted if Conditions 1, 2, 3, and 4 are all satisfied. These conditions are:

1. Number of members voting affirmatively must be more than that of members voting negatively (simple majority, abstention excluded).
2. There are at least 25 affirmative weighted votes.
3. There are at most 22 negative weighted votes.
4. There are at most 3 members voting negatively.

If any of the conditions are not satisfied, the votes of members from EEC countries shall be counted separately, and the proposal shall be adopted if Conditions 1, 2, 3, and 4 are all satisfied by these votes alone. The CEN standard is, in this last case, not formally binding for non-EEC countries.

The new approach on standardization calls for performance standards as opposed to descriptive standards, hence a greater need for methods for the measurement of performances and definition of thresholds and also a greater divergence from the usual definition of standards as an expression of a commonly recognized state of the art.

The working schedule of the standardization bodies is extremely heavy; at CEN alone, there are about 2500 projects dealt with by 250 technical committees and 1500 working groups.

There are three types of activities within CEN:

1. Mandates from the CEC for free circulation.
2. The Public Procurement Mandate.

TABLE 1—*Weighting of votes.*

Member Country	Weighting
France	10
Germany	10
Italy	10
United Kingdom	10
Spain	8
Austria	5
Belgium	5
Greece	5
Netherlands	5
Portugal	5
Sweden	5
Switzerland	5
Denmark	3
Finland	3
Ireland	3
Norway	3
Luxembourg	2
Iceland	1

3. Requests from European industry, e.g., from industrial federations which have developed industry standards and ask CEN to adopt them as European standards, or need new standards.

The work under the CEC mandate represents 70% of the CEN workload. The activities going on in the NDT field within TC 121 and TC 138 result from a CEC mandate of 1989 on the development of standards in the field of "simple pressure vessels."

Prenormative Activities

Framework Program

Some points of the framework program of the CEC research and development activities are devoted to prenormative work. To perform this, the Commission has direct or indirect action possibilities. Direct actions are performed by the Commission itself at its Joint Research Centre (JRC). Indirect actions, also called shared cost actions, result from agreements, contracts, etc. between the Commission and national industries and institutions. A few excerpts from the framework program of interest in the nondestructive testing sector are:

1. **Measurement and Testing Program.**
   A. *Direct actions:* some work on the reliability of structures.
   B. *Indirect actions:* Community Bureau of Reference (BCR). This Bureau, sometimes in collaboration with the JRC but mainly through shared cost actions, is involved with the creation of reference materials and with the definition and the harmonization of particular test methods.
2. **Materials and Industrial Technologies.**
   A. *Direct actions:* In the field of ceramics and composites at the JRC of Petten in the Netherlands, the CEC is, in its Institute for Advanced Materials, working on the development of advanced ceramics. In this field, at the CEN level, a specific technical committee has already been created (CEN/TC 184—Advanced Technical Ceramics); in parallel, the development of specific nondestructive testing techniques is underway at the JRC, Ispra, Italy.
   B. *Indirect actions:* Basic Research in Industrial Technology for Europe (BRITE); European Research in Advanced Materials (EURAM). Both programs are collaboration activities between multinational SMEs and the Commission. The outcome of these contracts is of high interest for the standardization bodies. Priorities are set with reference to the advice of IRDAC (Industrial Research and Development Advisory Committee of the CEC) and of European industrial needs.
3. **Reactor Safety Program (Part of the FISSION program).**
   A. *Direct action:* Program for the Inspection of Steel Components (PISC). This program has for many years been involved with the in service inspection (ISI) of nuclear pressure vessels and primary circuits by nondestructive testing means. The results of the second phase, PISC II, have already led to a modification of the ASME Code. The third and last phase, PISC III, is still going on, and here also output for codes and standards-making bodies will become available.
   B. *Indirect action:* DG XII. Working Group Code and Standards (WGCS). This working group is, in the framework of the Reactor Safety Program, aiming to create codes for the FBR as well as for the LWR.

CEC General Directorates (DGs)

Some committees or working groups of the DG III: Internal Market and Industrial Affairs, DG XI: Environment, DG XII: Science R&D, DG XIII: Telecommunications, and DG XVII: Energy are occasionally performing or commissioning prenormative work. IRDAC, the Commission's Advisory Committee on Industrial Research and Development, is making recommendations on the volume, the type of work, and the priorities for standards-making bodies, and advises the Commission on the prenormative work that should be performed within the different community programs.

European Conformity Assessment Systems

Although CEN has set up a framework for European certification, known as CENCER, allowing a European mark to be issued in respect to goods satisfying standards and for mutual recognition of test results, it was felt that some further steps were needed. Therefore, a new organization at the European level, called European Organization for Testing and Certification (EOTC), has been established in order to provide a common ground to negotiate mutual recognition agreements and to discuss common problems in the field of conformity assessment. EOTC should create an environment where mutual understanding and confidence can grow, a prerequisite to success in the field of voluntary certification.

Achievements to Date

European Standards of the 2900 and 45000 Series

Prior to making standards for nondestructive testing, European harmonization standards were developed in the field of quality assurance (five standards: EN29000 to EN29004) and in the field of testing laboratories, accreditation bodies, and certification (seven standards EN45001 to EN45003 and EN45011 to EN45014).

For the quality assurance standards (EN29000 series), the International ISO9000 Series Standards were adopted without modifications. For the testing laboratories accreditation bodies and certification (EN45000 series), the European standards were developed mainly by taking the existing ISO/CEN guide documents into consideration.

NDE Standards

Today, nondestructive testing standards are mainly developed within two technical committees of CEN: TC 138 and TC 121.

1. *TC 138—Non-Destructive Testing,* Secretariat France, was created in 1988 and after a short time, the following working groups actively started their work:

WG1: Ionizing Radiations	Secretariat	Germany
WG2: Ultrasonics	Secretariat	Denmark
WG3: Eddy Currents	Secretariat	France
WG4: Liquid Penetrant Testing	Secretariat	Germany
WG5: Magnetic Particle Testing	Secretariat	Spain
WG6: Leak Testing	Secretariat	Italy
WG7: Acoustic Emission	Secretariat	Italy

TC 138 has a mandate to develop generic standards on terminology equipment characterization and general principles for the different NDT techniques and also to prepare a document on the certification of NDT personnel. For the latter, no working group has been created due to the decision of the TC to handle this task itself.

2. *TC 121—Welding,* Secretariat Denmark, was created in 1987, but it was not until 1989 that an activity in the NDT field was started. In fact, the TC 121, Working Group 5—Inspection and Testing, was subdivided in 1989 into Subworking Groups: A: Destructive Testing and B: Non-Destructive Testing.

This WG5B, with France as secretariat, was organized as follows:

SWG1: Ionizing Radiations	Secretariat	Germany
SWG2: Ultrasonics	Secretariat	Denmark
SWG3: Eddy Currents	Secretariat	France
SWG4: Liquid Penetrant Testing	Secretariat	Germany
SWG5: Magnetic Particle Testing	Secretariat	Finland
SWG6: Visual Testing	Secretariat	United Kingdom

The mandate of TC 121 is to develop application standards for the NDT examination of welded structures including acceptance criteria.

It is a matter of fact that the first batch of standards in both committees is oriented to those standards that have been declared by industry to be the most urgently needed and have a widespread application in traditional production. However, a further goal is to provide a complete set of CEN standards also for new products manufacturing processes.

Ongoing Work within TC 138. Non-Destructive Testing

A. **Qualification and Certification of NDT Personnel**
At the very first meeting of the TC in January 1989 the decision was made to use ISO/DP 9712-3 as a basic working document, and after two years of work (end of 1990) the document "Qualification and Certification of NDT Personnel: General Principles" was ready for the six-month inquiry procedure. A further document entitled "Rules of Application of the General Principles to Complete the Draft on the Qualification and Certification of NDT Personnel" has been placed in the working program of TC 138.

B. **WG1: Ionizing Radiations:** Work items and Progress of Work
1. Terminology: ongoing work on the basis of ISO/DP 5576.
2. General Principles: finalized and under inquiry since mid-1990.
3. Equipment:

	Target date
IQI—Part one: wire type—finalized	06.1990
IQI—Part two: step hole type—finalized	09.1990
IQI—Part three: IQI values for steel and aluminum	09.1991
IQI—Part four: Experimental evaluation of image quality values and image quality tables	09.1991
IQI—Part five: Duplex IQI type	09.1991
4. Film system classification	09.1991
5. Tube voltage evaluation for X ray sources up to 450 KV (energy measurement)	09.1991
6. Test methods for the measurement of the effective focal spot size	09.1991

C. **WG2: Ultrasonics;** Work Items and Progress of Work

1. Terminology document finalized and under inquiry	12.1990
2. Calibration block; ISO ready for inquiry	12.1990
3. Calibration block; DIN 54120 ready for inquiry	12.1990
4. General Principles of Ultrasonic Testing	
Part 1: Generalities; ready for inquiry	12.1991
Part 2: Sensitivity	09.1991
Part 3: Transmission technique	09.1991
Part 4: Transfer correction	12.1991
Part 5: Geometrical conditions	12.1991
Part 6: Tandem examination	09.1991
Part 7: Characterization and sizing of discontinuities	12.1991
5. Characterization and verification of equipment	11.1992

D. **WG3: Eddy currents;** Work items and Progress of Work

1. Terminology	12.1991
2. General principles	12.1992
3. Characterization of equipment	12.1992
4. Defect parameters detection	12.1992
interpretation	12.1992
false calls	12.1992

E. **WG4: Liquid Penetrant Testing;** Work Items and Progress of Work

1. Terminology	03.1991
2. General principles; finalized and under inquiry	12.1990
3. Equipment	
4. Reference blocks	09.1991
5. Characterization and verification of products	03.1992
6. Function test	09.1991
7. PT equipment	06.1991
8. UV light sources	03.1991

F. **WG5: Magnetic Particle Testing;** Work Items and Progress of Work

1. Terminology	03.1991
2. General principles	09.1991
3. Magnetic particles: product and media	09.1992
4. Magnetizing equipment for MP flaw detection	03.1992

G. **WG6: Leak Testing;** Work Items and Progress of Work

1. Terminology	03.1991
2. Guide to the selection of a LT method	12.1991
3. Pressure test methods	06.1992
4. Tracer gas	04.1993

Ongoing Work within TC 121-5B Nondestructive Testing of Welds

A. **SWG1: Ionizing Radiations;** Work Items and Progress of Work

1. X-ray examination of fusion welded joints; finalized and under inquiry since June 1990	
2. Acceptance criteria	12.1991

B. **SWG2: Ultrasonics;** Work Items and Progress of Work

1. Ultrasonic examination of welds; finalized	12.1990
2. Acceptance levels using simple methods	12.1992
3. Acceptance levels using defect characterization	12.1992

C. **SWG3: Eddy Current;** Work Items and Progress of Work
   1. Inspection of welds by the EC method; finalized   12.1990
D. **SWG4: Liquid Penetrant Testing;** Work Items and Progress of Work
   1. General principles; finalized and ready for inquiry   12.1990
   2. Testing of PT materials   12.1991
   3. Reference blocks   12.1991
   4. UVA light sources   12.1991
   5. Equipment   12.1992
E. **SWG5: Magnetic Particle Testing;** Work Items and Progress of Work
   1. MP Testing of welds; finalized and under inquiry   06.1990
   2. Acceptance criteria   12.1991
F. **SWG6: Visual examination;** Work Items and Progress of Work
   1. Visual examination of fusion welded joints; finalized and under inquiry since end of 1990.

Conclusions

The European internal market has to be created in particular by the removal of technical barriers to trade. It is also important to secure European cohesion in particular in the field of emerging technologies by the creation of a common industrial competitiveness.

The adoption of the Single European Act and the new approach to technical harmonization have allowed, over a very short period of time, the adoption of EC directives covering a large range of industrial products and specifying only essential safety requirements, leaving it up to European standardization bodies to establish the standards, which are necessary for the effective implementation of the directives.

European standards are going beyond the safety requirements of directives. They introduce such aspects as effectiveness and reliability, elements of great economic importance.

European standards are to be adopted in all industrial sectors. Several are in the process of being elaborated and many more should be available by January 1993.

The Commission is quite aware of the challenge and of the problems facing European standardization and recommends that industry "give standardization a much higher priority in its strategy for the internal market."

Without greater involvement of industry in the work, the ambitious goals set by the Commission, CEN, and CENELEC may not be met. Lack of involvement at a strategic level by European industry is likely to be a high-cost option and will reduce the potential of the internal market.

References

[1] "Completing the Internal Market," white paper from the Commission to the European Council (Milan, 28–29 June 1985) COM(85)310 final, Brussels, Belgium, 14 June 1985.
[2] "Commission Green Paper on The Development of European Standardization: Action for Faster Technological Integration in Europe," 08.10.1990 COM(90)456 final, Brussels, Belgium.

J. C. Spanner[1]

Some Thoughts to Stimulate Discussion*[2]

REFERENCE: Spanner, J. C., "**Some Thoughts to Stimulate Discussion,**" *Nondestructive Testing Standards—Present and Future, ASTM STP 1151,* H. Berger and L. Mordfin, Eds., American Society for Testing and Materials, Philadelphia, 1992, pp. 180–182.

KEY WORDS: standards, standards writing groups, standardization of technology

Your attendance at this symposium suggests that you have unique interests within the NDT technology. Many of our colleagues probably consider our enthusiastic involvement in the standards writing process as strange—perhaps even a symptom of mental derangement. I know that most of my associates at Battelle who are actively engaged in research and development work cannot understand why I am so committed to standards writing when I could be in the laboratory doing something "useful."

It requires a variety of skills to achieve meaningful progress in this technology and I thoroughly enjoy, and highly value, my associations with people like you. Many of the best friends I have are members of the E07 Committee on NDT. The following comments reflect some of my philosophical views on the industrial standards writing process and I welcome this opportunity to share them with you.

The maturity of any given technology can rarely be precisely defined. However, relative indicators such as professional society activities; major technical symposia; the availability of equipment, services, and training; and the *publication of industrial standards* seem to be among the more relevant factors.

Webster's Unabridged Dictionary defines the term *standardize* as follows: ". . . to adopt a specified method as the only one to be utilized." This definition implies that whenever a group attempts to *standardize* a particular NDT method, sufficient development and application experience is available to support the requirements that are specified. It is appropriate to consider the consequences of this implication.

First, it is obvious to most practicing engineers that the process of standardization tends to encourage wider industrial usage of a given technology. Second, and perhaps less obvious, is the consequence that the standardization process tends to "freeze" the technology at the level specified in the standard.

If you question the second statement, consider what happens when a bidder responds to a bid package that requires compliance with an industrial standard, and the bidder offers a technical approach that is not consistent with that standard? Such bids are often automatically

---

* Partially extracted from "An Overview of Acoustic Emission Codes and Standards," by J. C. Spanner, Sr., *Journal of Acoustic Emission,* Vol. 6, No. 2.
[1] Staff engineer, Pacific Northwest Laboratory (operated by Battelle Memorial Institute), P.O. Box 999, Richland, WA 99352.
[2] Instead of the scheduled panel discussion on the role of NDT standards in world trade, a less structured, free-form discussion was conducted to provide a forum for the seminar attendees to exchange thoughts on this subject. S. J. Lavender, E. E. Borloo, M. Stadthaus, J. C. Spanner, and others offered remarks for consideration by the attendees. Mr. Spanner's remarks are reproduced here.

disqualified. The burden of proof for showing that the method offered is at least equivalent to the *standard* method rests solely with the bidder. The buyer may often exercise total discretion in accepting or rejecting such bids; even if the technology offered is clearly superior to the *standard* technology specified in the bid package.

Serious consideration of the preceding suggests that standards-writing groups should be guided by judgmental criteria, as well as technical criteria. Such judgmental criteria are difficult to quantify, but I suggest that the following merit consideration:

1. Standardization is appropriate only when the industrial need is sufficient to justify the effort.
2. A standard should not be written until an adequate technology and experience base is available.
3. Standards should only be developed by groups with a balanced representation of users and suppliers. (In this context, regulatory and similar interests are considered to be users.)
4. Industrial standards should emphasize the buyer's interests since the seller intrinsically possesses a greater knowledge of the technology.
5. No industrial standard should be based on the exclusive use of proprietary methods or techniques.

In my experience, not all standards committee members recognize or fully appreciate the importance of such criteria. On occasion, I have observed poorly attended committee meetings where most of those present represented supplier interests. When such meetings are either poorly attended or unbalanced with respect to user and supplier interests, the basic strength of the consensus process is jeopardized.

In the economic climate of the past few years, it seems to me that reduced participation in codes and standards activities has been more evident among the users than the suppliers. This simply doesn't make sense and is another example of the shortsightedness that U.S. industry and many government agencies have exhibited for the past decade or more. It seems imperative that the users recognize and accept their responsibilities for protecting themselves and the public. And I commend the suppliers for their continued willingness to serve on such committees, even during the tough times.

Critical Balance Point

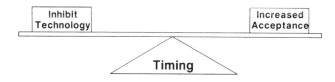

• Technology Base   • Market Size and Need
• Experience Base   • Equipment Availability
• Technology Forecast   • Operator Availability
• Economic Aspects   • Foreign Competition

FIG. 1—*Key factors to consider when initiating work on a new standard.*

Critical Balance Point

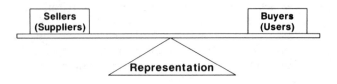

- ● Profit Potential
- ● Marketing Aid
- ● Technical Advantage
- ● Knowledge Advantage

- ● User Interests (Technical)
- ● Public Interests (Safety)
- ● Assure Competition
- ● Knowledge Disadvantage

FIG. 2—*Key factors to consider when selecting committee members.*

Whenever work is initiated on a new standard, the committee should carefully assess:

1. The true need for the standard.
2. The available technology and experience base.
3. The need to maintain a balance of interests on the committee.

These considerations are depicted in Figs. 1 and 2 where the concept of a *critical balance point* is used to illustrate the key factors that should guide and influence any standards writing group.

It is also appropriate to consider whether the technology is sufficiently strong to drag the anchor that will be attached to it when the standard is published. If not strong enough, the technology may not advance as it might have without the standard.

In conclusion, it seems to me that standards writing groups should be guided by judgmental, as well as technical, criteria. Whenever work is initiated on a new standard, the committee should carefully consider: (1) the need for the standard, (2) the available technology and experience base, and (3) the need to maintain a balance of interests on the committee. In addition, the technology should be sufficiently advanced to withstand the inhibitions that standardization will impose on future development.

NDT Personnel Qualification: Here and Abroad

George C. Wheeler[1]

NDT Personnel Qualification—A U.S. Perspective

REFERENCE: Wheeler, G. C., "NDT Personnel Qualification—A U. S. Perspective," *Nondestructive Testing Standards—Present and Future, ASTM STP 1151,* H. Berger and L. Mordfin, Eds., American Society for Testing and Materials, Philadelphia, 1992, pp. 185–194.

ABSTRACT: The status of NDT personnel qualification and certification (PQ) is reviewed from the point of view of U.S. practitioners. Present U.S. practices and standards are compared with those of some other nations and those of ISO.
  Results of a survey of U.S. organizations and individuals familiar with various PQ systems are used to evaluate the spectrum of current U.S. attitudes on the subject. It is concluded that an international standard acceptable to the United States as well as to most other nations should now be achievable in a relatively short time.

KEY WORDS: personnel qualification, personnel certification, PQ, U.S. attitudes, PQ practices, PQ standards, future directions

Standards for NDT personnel qualification and certification (PQ) vary widely among U.S. NDT organizations and among nations. It is thus difficult for a purchaser to determine whether a supplier's personnel are sufficiently competent for the required NDT work. The results are often uncertainty, additional testing, multiple certifications, and excessive auditing. The impediment to trade is obvious, yet few will deny the need for proper qualification of NDT personnel. NDT work is relatively expensive and strongly operator dependent, often subjective, and frequently crucial to safe or satisfactory operation of costly or hazardous equipment and structures.

The American Society for Nondestructive Testing (ASNT) has in recent years taken several significant steps toward improving the situation and achieving harmonization with other nations. It has supported the work of ISO TC135/SC7 on PQ, both financially and technically. In recent months, an ASNT committee has been formed expressly to determine the U.S. consensus position on PQ documents promulgated by ISO TC135. Perhaps most important, the ASNT Standard for Qualification and Certification of Nondestructive Testing Personnel (ASNT CP-189) was developed and published embodying many important changes designed to overcome the weaknesses of current systems.

On 15 March 1991, ASNT CP-189 was approved by the American National Standards Institute (ANSI) as American National Standard ANSI/ASNT CP-189-1991.

The major innovations introduced in the ANSI/ASNT standard are:

1. All Level III personnel must be centrally certified by ASNT in addition to certification by their employers. The ASNT certification, administered by a technically expert certifying board, assures a uniform, high minimum level of knowledge. Competent and eth-

[1] President, Wheeler Nondestructive Testing, Inc., 29 Front St., Schenectady, NY 12305.

ical performance after certification is encouraged by a strict code of ethics backed by strong, actively imposed sanctions.

2. The prerequisite experience for Levels I and II is essentially double the minimum guidelines of ASNT Recommended Practice for Personnel Qualification and Certification in Nondestructive Testing (SNT-TC-1A), which is the basis of most current U.S. certifications. Further, the "twenty-five percent rule" of SNT-TC-1A has been eliminated and, for the first time in any PQ specification, requirements are established for instructors.

3. Level III is made responsible for all aspects of training, examinations, certification, and subsequent performance of the employer's NDT personnel. The intent is to strengthen the position of Level III within the employer's organization, thereby improving the quality of the NDT work performed and reducing the ability of any unscrupulous employer to improperly influence NDT work.

4. All personnel are prohibited from becoming certified by passing examinations which they, themselves, prepared.

Although the ANSI/ASNT standard has reduced the differences in PQ practices between the United States and other nations, some differences still exist and some new ones have been created in the effort to close loopholes and strengthen requirements. What, then, can be done to reduce or eliminate the differences among certification practices in order to promote trade?

Major Differences Among PQ Systems

Systems Studied

In considering what can be done, it is first necessary to establish what the fundamental differences are among PQ systems. An analysis was made of thirteen existing and proposed PQ standards, including eight from countries or organizations outside the United States. The documents studied were:

1. ASNT Recommended Practice SNT-TC-1A (SNT-TC-1A).
2. ASNT/ANSI Standard for Qualification and Certification of Nondestructive Testing Personnel, ANSI/ASNT-CP-189 (ANSI/ASNT standard).
3. U.S. Navy Welding Standard (NAVSEA 250-1500-1).
4. U.S. Navy Requirements for Nondestructive Testing Methods (MIL-STD-271).
5. U.S. Air Force NDT Personnel Qualification and Certification (MIL-STD-410).
6. ISO Draft International Standard for Qualification and Certification of NDT Personnel, ISO-DIS-9712.2 (DIS 9712).
7. Nordtest Scheme for Examination of Nondestructive Testing Personnel (Nordtest).
8. Certification of NDT Personnel in Britain (PCN).
9. Certification Scheme for Weldment Inspection Personnel (CSWIP).
10. JSNDT Rules for Certification of Nondestructive Testing Personnel (NDIS 1601).
11. PRC National Standard for Qualification and Certification of Personnel (GB-9445).
12. DGZfP Recommendation for Qualification and Certification of NDT Personnel (DGZfP).
13. CICPnD Standard for NDT Personnel Certification (ST-1).

Most Serious Differences

Although there are many differences among the systems, they are primarily in details. For example, virtually all systems have now adopted three levels of certification, as pioneered by

SNT-TC-1A in 1965. However, some assign different names to the levels, some add another level or two, and a number differ in the activities permitted for some levels of personnel. Similar differences of detail exist in prerequisites for examination, the types and content of examinations, methods of grading, and so forth. None of these appear to constitute a major barrier to general mutual acceptance of certifications among nations and organizations.

The differences which have caused or seem likely to cause serious problems may be broadly classified into four categories.

1. The characteristics and power of the certifying body, and requirements for membership therein.
2. The prerequisites for certification, including the qualifications of instructors.
3. The mandatory responsibilities, if any, that are assigned to certified personnel to assist in securing competent performance.
4. The provisions, if any, that are included to enforce ethical conduct of certified personnel.

The Certifying Body—Provisions regarding the certifying body are among the most contentious differences among the systems. Most national systems, except those used in the United States, require certification of all levels by a central body not associated with the candidate's employer. In the United States, employer certification of all levels is the rule with the exception of NAVSEA 250-1500-1 and the ANSI/ASNT standard. These standards require central certification of Level III personnel, with employer certification of Levels I and II; additionally, the ANSI/ASNT standard mandates employer specific examinations (and in some situations, employer practical examinations) for Level III personnel and employer certification of all levels.

Many of the non-U.S. specifications also require that the certifying body may not provide any of the prerequisite training or education of the candidates.

A third major difference lies in the criteria for members of the certifying body. The ANSI/ASNT Standard, by specifying that all Level IIIs be certified by ASNT, has insured, albeit indirectly, that the certifying body will be technically competent. This is a result of ASNT's stringent written requirements for membership on its National Certification Board. No other system studied provides for assured technical competence of the certifying body.

Prerequisites for Certification—All systems require candidates for certification to possess some degree of education, training, and/or experience in NDT prior to taking certification examinations. Again, while there are differences in such details as the hours of training or months of experience, the differences are seldom so large as to present a significant problem in achieving recognition of certification equivalence between systems.

A more difficult issue exists between systems where one of these prerequisites is omitted in one system while another emphasizes it. For example, ISO-DIS-9712 has no training or education requirements for Level III, while the ANSI/ASNT standard ties the experience requirements to the degree of formal education which the candidate has achieved. The Japanese standard, JSNDT NDIS-1601, on the other hand, has no training requirements for Level II, but does for the other levels; it also has four levels, rather than three.

The ANSI/ASNT standard also contains prerequisites for NDT Instructors, a desirable requirement not contained in any other system.

Mandatory Responsibilities—One of the weaknesses often criticized in certification practices is the undesirable influence which may be exercised by an ignorant or unethical employer when the principal NDT expert does not have sufficient power or fortitude to assure proper performance of the NDT activities. Except for the Nordtest scheme and the ANSI/ASNT standard, none of the systems addresses this problem. The ANSI/ASNT standard does so by assigning broad, mandatory responsibilities to the Level III, backed up by sanctions which will impact the employer as well as the Level III. Nordtest provides that Level III job functions and responsibilities shall be specified in detail in a written employment contract.

Enforcement of Ethical Conduct—Most systems contain provisions, either implicit or explicit, requiring ethical conduct by candidates as well as by certified personnel. The ANSI/ASNT standard goes further than others in two respects. First, it prohibits any individual from taking an examination which that individual prepared. Second, through the requirements of the ASNT Level III Certification Program, it imposes a stringent Code of Ethics on the Level III, backed up by a formal procedure for sanctions which may include loss of certification. Since ASNT certification is required for all Level III personnel under the ANSI/ASNT standard, any sanctions imposed will also affect the certification status of the employer's other NDT personnel and thereby affect the employer.

Reasons for U.S. Differences

Survey of U.S. Industry PQ Practices

To establish a factual basis for discussion of why U.S. practices vary so significantly from those of most other nations, a survey was undertaken. Seventy U.S. experts on NDT personnel certification representing a cross section of the NDT industry were asked to complete a four-page questionnaire regarding facts, beliefs, and attitudes about PQ which are prevalent in their segment of the industry. In terms of the usual consensus categories, 21 of those canvassed were NDT users, 15 were experts, 28 were suppliers, 3 were labor, and 3 were regulatory bodies. Manufacturers, fabricators, constructors, technical societies, metal producers, and service organizations were among those canvassed.

Twenty-four individuals, 34% of the total, responded to the survey. By consensus category, 20% of the experts responded, 38% of the users, and 46% of the suppliers, but none of the labor or regulatory bodies. These results are summarized in Table 1. The responses came from the following industrial segments:

1. Aerospace.
2. Nuclear equipment.
3. Chemical.
4. Nuclear services.
5. Construction.
6. Petroleum equipment.
7. Electrical and electronics.
8. Pressure vessel.
9. Fabrication.

TABLE 1—*Experts canvassed and responding to survey.*

Canvassees		Respondents	
No.	Category	Number	Category, %
21	Users	8	38
15	Experts	3	20
28	Suppliers	13	46
3	Labor	0	0
3	Regulatory	0	0
70	Total	24	34

10. NDT services.
11. Metals producing.
12. Shipbuilding.
13. NDT equipment and materials.
14. Utilities.

Analysis—PQ Systems Used

As shown in Table 2, 46% of the respondents reported that their segment of the industry had adopted a single system for NDT personnel qualification, while the balance stated that more than one system was required in their industry. Of those with a single system, 91% were using employer certification under SNT- TC-1A and 9% were using MIL-Std-410. Since both of these documents are based on employer certification, 46% of the organizations responding are operating totally under employer certification systems.

The dominance of employer certification in the United States is further emphasized by the responses of those using more than one system. All of the group use employer certification in some form for at least part of their work. The breakdown by system is that 92% use SNT-TC-1A, 69% use MIL-Std-410, 54% use MIL-Std-271, 31% use NAVSEA 250-1500-1, 15% use the ASNT Standard, and 8% use the EPRI certifications for IGSCC. Of these systems, only the latter three involve any central certification, and the EPRI certification is confined to one application of one NDT method.

Analysis—Importance of International Trade

An attempt was made to assess the importance of international trade to the respondents' organizations. Fifty-eight percent indicated that their organization makes a strong effort to market its products or services in other countries. All the industrial countries and most of the developing nations were cited as customers. Only four respondents stated that any overseas customer required them to use a PQ system other than their own normal practice, and in every case other respondents reported different experience with the same customer nations.

Analysis—Cost of Additional Certifications

The most frequently used argument heard in the United States against central certification is that it will greatly increase the cost of performing NDT work. An attempt was made to elicit quantitative data on the issue by asking for an estimate of the additional annual costs incurred per certified person for each additional PQ system employed. Ten answers were received, ranging from $180 to $4000/year/person/system, with a median of $500 and a mean of $1185.

TABLE 2—*PQ systems used in the United States.*

Situation	System	Percent
Only one system used (46%)	SNT-TC-1A	91
	MIL-STD-410	9
Several systems used (54%)	SNT-TC-1A	92
	MIL-STD-410	69
	MIL-STD-271	54
	NAVSEA-250-1500-1	31
	ANSI/ASNT Std	15
	EPRI IGSCC	8

TABLE 3—*Preferred PQ system for suppliers to use.*

System	Preference (Most⟶Least) 1	2	3	4
ANSI/ASNT Standard	88%	4%		
ISO-DIS	4%	13%	4%	
Nordtest		4%		4%
CSWIP		9%	4%	4%
JSNDI	4%			

Analysis—Preferred Systems for Central Certification

In an effort to obtain evaluations of the better known central certification standards, canvassees were asked to choose which system they would prefer (1) if central certification were required of them and (2) if they were to impose central certification upon their suppliers, together with the reasons for their choices.

Preferred for Suppliers—The system most would prefer to have their suppliers use was the ANSI/ASNT standard, 88% selecting it as their first choice, and 4% as second choice. A wide variety of reasons were listed, but the majority mentioned familiarity or similarity to present practices; the next largest response, 17%, cited the assurance that all Level III personnel would be ASNT-certified.

The second choice among systems was ISO-DIS-9712, 4% naming it as No. 1, 12% giving it second place, and 4% assigning it to third place. The global nature of their company's business was the factor most important to 80% of the respondents.

CSWIP and Nordtest were the only others mentioned more than once, and JSNDI was the only one, other than ASNT and ISO, to be named as first choice. Table 3 summarizes these results.

Preferred for Own Company—If central certification in any form were required of their own company, 88% would prefer to work to the ASNT standard, with 79% naming it as their first choice and 9% as second choice, as shown in Table 4. Familiarity was given as the reason by 38% of the canvassees, while 17% named cost and another 17% considered it technically superior.

ISO-DIS-9712 was the second most popular system, being preferred by 17% of respondents, with 9% choosing it as No. 1 and 4% each selecting it as second and third choice. Global business and a desire to avoid use of more than one system were mentioned as reasons for the choice.

TABLE 4—*Preferred PQ system for own company to use.*

System	Preference (Most⟶Least)			
ANSI/ASNT Std	79%	9%		
ISO-DIS	9%	4%	4%	4%
Nordtest		4%		4%
CSWIP		9%		4%
JSNDI	4%			

TABLE 5—*Valid reasons for employer certification.*

Reason	% of Respondents
Lower cost	58
Meet employer needs	54
None are valid	21
None for Level III	9
Familiarity	4
Preferred by union	4
Lower cost for Levels I and II	4
Requires less time	4

Analysis—Factors Favoring Employer Certification

Canvassees were asked what they consider to be valid arguments favoring employer certification as opposed to central certification. The most frequently mentioned (58%) was cost, followed closely (54%) by the desirability or need to tailor the qualifications to the employer's business, as shown in Table 5.

Twenty-one percent of the respondents feel there is no valid reason to prefer employer certifications, while nine percent stated this is the case for Level IIIs (presumably implying that valid reasons exist for Level I and II). Union preference, less time required, flexibility, and objection to specific and practical examinations being given centrally made up the balance of the replies.

Analysis—Factors Favoring Central Certification

Improved quality and performance of NDT was cited most often (58%) as a valid argument favoring central certification as opposed to employer certification. Greater uniformity of personnel quality was mentioned by 41%, and 30% felt that central certification would improve the credibility of NDT personnel and results. Thirteen percent felt that costs would eventually be lower with central certification, while 9% each mentioned portability of the certificates, objectivity of the certifications, improved ethics of personnel, and a reduction in number of audits resulting from use of a single system by everyone. Nine percent feel there are no valid arguments for central certification. Table 6 summarizes these responses.

TABLE 6—*Valid reasons for central certification.*

Reason	% of Respondents
Better NDT quality/performance	58
More uniform personnel quality	41
Improved NDT credibility	30
Lower long-term costs	13
Portable certifications	8
More objective certifications	8
Improved ethics in NDT	8
Reduction in no. of audits	8
None are valid	8

TABLE 7—*Who should be centrally certified?*

	Responses		
Level	All	Some	None
Level III	63%	25%	8%
Level II	29%	17%	46%
Level I	17%	4%	71%

Analysis—Who Should Be Centrally Certified?

To further explore the attitudes concerning central versus employer certification, canvassees were asked whether their industry feels central certification is desirable for all, some, or none of their personnel in each Level. The results are shown in Table 7. For Level IIIs, 68% answered "all," 27% said "some," and 4% responded "none." This surprising result means that 95% feel that at least some of their Level III personnel should be centrally certified.

As expected, there was less enthusiasm for central certification of Level II. Thirty-two percent favored it for all Level IIs, 18% said "some," and 50% were opposed to it for any Level II. Surprisingly, as many favored central certification for at least some Level IIs as were opposed.

With regard to Level I personnel, 18% favored all being centrally certified, 4% answered "some," and 77% were opposed to any central certification. The 22% favorable to such certification for at least some of their Level I personnel is about what was expected.

Analysis—Other Comments

Two questions on the survey invited comments on issues not covered by the other questions. One asked for items critical to the attitude of the respondent's industry regarding PQ issues, while the other solicited general comments. These were analyzed together because many of the responses overlapped the categories, while others dealt with factors previously covered. The unique answers provided were, in no particular order:

"... my [aerospace] industry ... [seems] to be totally unconcerned about PQ&C issues."

"... if [contracts] did not require SNT-TC-1A ... [the] construction [industry] would not be using [any system] ..."

"Certification is becoming a job security issue ... it's now ... used as another crutch for inefficiency."

"I fear bureaucracy [of central certification] ... [resulting in] unjustified costs ... lowest [common] denominator [quality] ... stifled originality."

"Our industry [materials producer] would embrace central certification only if the customer required it."

"NAVSEA [250-1500-1] has been very successful ... using central certification of the ... Level III ... [only]."

"The ISO ... [DIS-9712] ... does not deal with the issue of equivalent certification ... examinations. This is the critical issue ... internationally."

"Central certification ... does not verify ... ability to perform ... Only the employer ... can judge ... attitude, ability ..." "Non-governmental central certifications are preferable to government action."

Discussion

The survey results make it abundantly clear that, with few exceptions, the U.S. NDT industry has little experience with central certification of any kind, much less the totally central systems used in most other nations. Furthermore, the extensive U.S. experience with employer certification dates back more than twenty years. Under these circumstance, the U.S. resistance to central certification can hardly surprise anyone.

What is surprising is the extent of the recent change in U.S. attitudes. Perhaps as little as nine, certainly fifteen years ago, there was virtually no acknowledgment that employer certification might have some shortcomings or that central certification might have important advantages. The survey reported herein indicates that today a substantial majority feel that central certification would improve the quality of NDT performed, and large minorities feel that it would improve the credibility and uniformity of certifications. In this climate, there should be excellent opportunities for the United States to work successfully in the international NDT community to assist in designing a PQ standard which all countries can accept.

There appear to be several critical issues which should be addressed in such an effort. These are:

1. Provisions regarding the extent and limits of central and employer certifications.
2. Provisions to assure that the certifying bodies possess strong, well-documented technical qualifications.
3. Provisions to assure minimum prerequisites for education, training, and experience for all Levels, including instructors.
4. Provisions to assure adequate oversight of NDT operations by Level III personnel.
5. Provisions to assure that unsatisfactory technical or ethical performance will be corrected.

The first step in the process should be to agree on the critical issues, whether they are those listed above or others. Second, it will be necessary to agree on the basic principles around which a consensus solution of these issues may be achieved.

Details regarding the principles must then be agreed upon, and finally the lesser details of the standard must be negotiated.

Throughout the process, certain findings of this study should be addressed. Although U.S. opinion now recognizes that some degree of central certification will probably improve NDT performance, the degree must be limited. There is a strong conviction that the degree of centrality in ISO-DIS-9712 will be excessively costly without compensating benefits. There is also a strong conviction that employer specific and practical examinations are so essential to the satisfactory performance of NDT in most industries that there is little to be gained, and there may be a net loss, if such examinations are conducted centrally.

These reservations regarding central certification may be most applicable to large, heavily industrialized nations such as the U.S.S.R. and the United States. There are a great many people to certify, distances are large, industry is very diverse, and many employers use highly specialized NDT equipment which it would not be practical to duplicate in central certification facilities. Regardless of whether such considerations apply to other nations, it is nevertheless important that they be taken into account if a generally acceptable PQ standard is to be achieved.

Conclusions

The time appears right to pursue formulation of an international PQ standard which would be acceptable to all nations. ISO-DIS-9712 has many desirable features, but lacks some of the essentials such as technical qualifications for the certifying body. It is also too heavily oriented toward central certification. With modifications in these and a few other areas, changes which might well be completed within two years, ISO-DIS-9712 could become a document that would probably be embraced by most if not all of U.S. industry, as well as by those nations that accept it in its present form. This is a goal worthy of serious pursuit by the entire international community. It would have an important impact on the quality and cost of NDT performed throughout the world, as well as on international trade.

John H. Zirnhelt[1]

NDT Personnel Qualification—An ISO Perspective

REFERENCE: Zirnhelt, J. H., **"NDT Personnel Qualification—An ISO Perspective,"** *Nondestructive Testing Standards—Present and Future, ASTM STP 1151,* H. Berger and L. Mordfin, Eds., American Society for Testing and Materials, Philadelphia, 1992, pp. 195–200.

ABSTRACT: Subcommittee 7, "Personnel Qualification," of the International Organization for Standardization (ISO) Technical Committee 135, "Nondestructive Testing," has been working since 1983 on the development of an international standard for the qualification and certification of nondestructive testing personnel. The document is based on fundamental principles proposed by the International Committee for Nondestructive Testing (ICNDT). The draft international standard (DIS 9712) was circulated for letter ballot in early 1990. General international agreement on DIS 9712's key elements of central certification at three levels in individual nondestructive testing (NDT) methods has been reached. The document has earned the support of developing countries, who embraced it even in its earliest format as a model for new national standards, and who, through their regional organizations, contributed significantly to its development. As we enter 1992, the pressure for a universally acceptable standard which permits mobility of personnel, boasts broad general international acceptance, and maintains the high level of competence needed in NDT will lead to agreement on the international level. A comprehensive ISO standard for the qualification of NDT personnel will facilitate international trade in capital goods and services.

KEY WORDS: nondestructive tests, qualification, certification, international harmonization, standards

The participating (P) members of ISO Technical Committee TC135, "Nondestructive Testing," and the member bodies of ISO are currently responding to a letter ballot seeking approval of Draft International Standard (DIS) 9712-2 [1]. This DIS has been developed by TC135's Subcommittee 7, "Personnel Qualification," in meetings beginning in October 1983 and continuing through December 1988. DIS 9712-2 represents a fair consensus of the countries who participated in the process.

DIS 9712-2 is the result of debate, compromise, and resolution of significant differences of the initial positions held by individual delegates. This document's present stage is a tribute to the spirit of cooperation that the subcommittee has experienced since its inception.

Incentive for a country to participate in the development of an international standard varies with geography, level of industrial development, and time.

There has been a strong need for an international standard for the qualification and certification of nondestructive testing (NDT) personnel for 25 years and especially in the last 8 years. The movement toward the present effort is a result of national and regional economics with general recognition of global market needs.

[1] President, Charcas International Inc., 22-1050 Britannia Road East, Mississauga, Ontario L4W 4N9, Canada.

The Demand

One of the earliest calls for an international system for the qualification of NDT personnel came at the Fifth International Conference on NDT, held in 1967 in Montreal, after delegates heard from speakers who described existing national certification schemes in Canada, Japan, the United Kingdom, and the United States [2].

The need to control the human factors which strongly influence the reliability of any nondestructive test was recognized at that time and at virtually every NDT conference since. Over the next two decades, national certification schemes were developed in most industrialized countries to meet local political needs, to fit local industrial environments, and to categorize the outputs of existing educational and training systems, sometimes with no reference to parallel developments in other countries.

Some level of international harmonization was considered generally desirable and actively promoted by several bodies including the International Committee for NDT and the European Committee for NDT. It was only in the last decade, however, that action became a matter of strong international concern due to a general trend toward market globalization and the growth of regional interests, most notably in Europe and in the developing countries [3].

Developing the Consensus

The process of developing and approving an international standard is a consensus process, designed to ensure that the philosophy and technical content of a proposed document is thoroughly debated. By its very nature, the system solicits compromise, concession, and dilution to satisfy the often conflicting demands of a sufficient number of members to develop a generally acceptable document. With any standard, the objective is to produce a document that is meaningful and useful, allowing flexibility to meet national situations, yet without compromising basic principles.

In this respect, the development of DIS 9712-2 has been along a typical path of review and revision of working drafts, draft proposals, and draft international standards, with regular meetings to consider in detail documents added to the discussions and related comments.

During its business meeting in 1967, the International Committee for Nondestructive Testing (ICNDT) passed a recommendation to the International Organization for Standardization (ISO) that a new ISO technical committee be formed to deal with the subject of nondestructive testing [4]. When TC135 was formed in 1970, its secretariat was assigned to the United Kingdom. In 1974, the United States accepted responsibility for TC135, and several years later it was reassigned to the USSR. At the 2nd Plenary Meeting of TC135, held in Philadelphia in 1975, the creation of Subcommittee 7, "Personnel Qualification," was approved and the secretariat accepted by ASNT [5]. Prior to the plenary meeting of ISO TC135 held in Moscow in 1980, however, the United States announced it had no longer any interest in providing the secretariat for SC7, and the committee accepted a proposal that Canada take over this responsibility.

The first meeting of SC7 was held in Ottawa in 1983 amid considerable controversy surrounding the question of whether an ISO committee had the mandate to deal with the qualification of personnel. There was also opposition from the International Committee for NDT's Working Group on Training and Certification, which had begun discussions of its own on international harmonization. During the two days of discussion, however, a decision was reached to proceed with the development of an international standard for the qualification and certification of NDT personnel, beginning with the formal step of registering this task as the single work item for SC7.

Among the basic principles agreed upon at that meeting was that the subcommittee in its work would take into full consideration the concurrent and precursor work of ICNDT, recognizing that ICNDT's results already represented a crucial first step toward international agreement. Agreement was also reached on some basic issues; it was decided that the qualification system would be by NDT method, in three levels, and would involve examinations under the supervision of an independent national body.

The second meeting was held in Paris in February 1985, at which time two key documents were reviewed. The first was a summary developed by the secretariat of a dozen or so national certification schemes, outlining the common elements and indicating the key differences to be resolved. The second document was a draft from Canada which was roughly based on the American Society for Nondestructive Testing's SNT-TC-1A, but structured with central certification at all levels. The subcommittee appointed a four-member Working Group under the chairmanship of France, with members from Japan, Canada, and the International Atomic Energy Agency (IAEA), to develop a new draft based on the results of the Paris meeting and the documents tabled there.

The Working Group worked through the summer, then met in Philadelphia in October 1985. From this Philadelphia meeting it produced a further working draft which was circulated in two versions for consideration at the next meeting of SC7.

In May 1986, ISO TC135 SC7 held its third meeting in Milan. The key document under consideration was the working draft produced by its Working Group. It also considered recommendations prepared by the International Committee for NDT at its 1985 Las Vegas meeting [6]. These recommendations and the training guidelines developed by the United Nations Regional NDT Project for Latin America and the Caribbean as an IAEA Technical Document [7] were accepted as a basis for the training requirements to be defined in the standard.

The document resulting from the SC7 meeting in Milan (ISO TC135 SC7 Document N35-E) was circulated for letter ballot in late 1986 and was subsequently registered by the ISO Central Secretariat in Geneva as Draft Proposal 9712, "Nondestructive Testing—Qualification and Certification of Personnel," in February 1987. Negative comments were received by the subcommittee secretariat as a result of this ballot from only 4 of the 26 members, suggesting that substantial international agreement had already been reached on the document at that stage.

After wide circulation of the draft proposal, it was further discussed at an SC7 meeting in Philadelphia (November 1987) and reissued as Draft Proposal 9712-2. DP9712-2 was reviewed at another SC7 meeting in Kingston, Jamaica (December 1988), circulated to the subcommittee members in revised form, and submitted to the ISO secretariat for balloting as a draft international standard (DIS).

DIS 9712 was circulated to member bodies in late 1989. When voting closed on 7 March 1990, the DIS had been approved by 14 of the 18 participating or "P" members of TC135 (78%, or well over the required 66.7%), but 7 negative votes were recorded from 25 ISO member bodies. These seven negatives represented 28% of the members voting and exceeded the criterion of 25%. DIS 9712 was therefore not approved [8].

Many useful comments were received by the secretariat as a result of the 9712 ballot, some of which will improve the clarity of the document. For the 9712-2 revision, however, the most significant comments were those of the United States (ANSI), which indicated that their incorporation would change its negative vote to affirmative. On the other hand, some of the comments supporting negative votes from ISO members which had not participated in the development of the document and presumably were seeing it for the first time, ran counter to the principal compromises which formed the consensus basis of the document.

It was anticipated that the results of the current ballot would be available by early in the third quarter of 1991, and, if approved, the International Standard would be published in late 1991.

The Role of the "A" Liaison Members of ISO

ISO procedures stress the desirability of liaison between technical committees and their sub-committees and international or broadly based regional organizations which share an interest in the work of the committees and indicate that the views of such organizations be taken into account at an early stage of the work.

The development of DIS 9712-2 has benefitted from continuing contributions from two international organizations which have registered as "A" category liaison members of SC7. Both organizations have been kept advised of the subcommittee's progress and both have sent representatives to SC7 meetings.

The International Committee for Nondestructive Testing, an international "federation" of national NDT societies, has provided a forum for widespread discussion on many occasions, and its Working Group on Training, Qualification and Certification contributed substantially to the groundwork of the DIS. The subcommittee therefore began its task in 1983 with some substantial consensus already established. The European Committee for Non-Destructive Testing also had an active Working Group whose views were contributed through the ICNDT.

The International Atomic Energy Agency became involved with the work of SC7 in 1985 following the recommendations of a group of consultants engaged to advise the Agency's Regional NDT Projects in Latin America and Asia on the development of a regional standard. The consultants' recommendation was that the Agency should monitor and contribute to the work of ISO TC135 SC7, rather than expend resources on repeating work of others [9]. IAEA has been a regular participant in the work of SC7, has provided tangible input to the standard in the form of its training guidelines [7], and has provided many opportunities for open and detailed discussion within the regional NDT projects on the subject of qualification and certification. These discussions have led to strong support among the developing countries for the DIS since an ISO standard is relatively easy for a country with no existing standard to adopt with little or no revision. Similarly, pressures for regional harmonization were forestalled in favor of international harmonization.

Europe 1992

One of the accelerating factors in the development of this standard has been the European Community's commitment to harmonization on a regional level by 1992. In essence, the participants in the development of European (CEN) standards have agreed that upon the issue of a CEN standard, national standards on the same subject will be withdrawn. Where a CEN standard is referenced in a European Community directive, its use in activities regulated by that directive becomes mandatory [10]. The pressure on SC7, then, has been to develop an ISO standard which could be adopted in essence by CEN.

National Implications of an ISO Standard

ISO Standards are the product of an international consensus system and, as such, are built from the contributions of individual members, each representing their respective national positions. These national positions are presumed to reflect, in turn, a consensus of various industries, sectors, and political organizations within each country.

There is no guarantee when a country participates in the development of an international standard that the final product will be fully acceptable to that country. Similarly, there is no obligation on any country, no matter how it votes, to adopt any ISO standard.

An ISO standard is sometimes adopted as a national standard in full simply by following the local standards writing and approval procedure and adding a new number to the ISO number. The ISO standard may be treated as a source or reference document and incorporated in part or in whole in a more comprehensive national document. The national standards writing committee may decide that the ISO standard is not sufficiently stringent and build a new standard by adding conditions.

An ISO Standard for Qualification and Certification of NDT Personnel

The DIS for qualification and certification of NDT personnel has been supported during its development by several diverse groups. The developing countries and newly industrialized countries have supported it because they see a need for a national standard, ideally one which is harmonized with a broad range of their trading partners. Simple adoption or adoption with minor modification has been the route of many countries including, among others, the Philippines, Thailand, Malaysia, Trinidad and Tobago, Jamaica, and Bolivia, which have used the ISO document as a model despite its draft status. Most of these countries had no existing national standard. Other countries, such as Argentina, Brazil, Australia, and Japan, have undertaken major overhauls of their existing standards to bring them in line with the ISO draft. In Europe, the ISO document is being used as the model for the CEN standard. In the United Kingdom, it has been interesting to watch the ISO model used as a framework to integrate several different industry-specific schemes under one national system. In Canada, the standards writing committee is still dealing with the changes it must make if its standards are to be harmonized with ISO.

On some issues, the SC7 members deliberately left flexibility for local conditions. For example, formal education, while recognized as contributing to the capability of an NDT operator to perform both on the job and in a practical examination, was impossible to define in a general manner that would apply to all countries. The need for proof of education was left (in clause 6.2 of the DIS) as something that "may be required to establish the eligibility of a candidate." Most national standards replace this clause with a specific statement generally reducing experience or training hour requirements as the education level increases.

Another example of local variation is in the table of experience requirements for Level 3 (Table 3 in DIS 9712-2). Many countries have included in their national standards a clause which makes a first degree in engineering or science a *requirement* for Level 3 certification.

Some Contentious Issues

From the beginning of discussions, the issue of central certification at three levels has been a point of contention, yet the clear majority of members insist on this as a critical and inviolable condition. There seems to be no serious concern about whether the national certifying body is an NDT society (particularly popular with ICNDT members), a government agency, or the national standards organization. On the other hand, there was insistence on the fact that there should only be one certification body in a country and that this body should somehow be recognized by the ISO member body. The national certifying body must also include qualified representation from all sectors within the country's NDT community; in other words, its administrative committee must be a consensus committee.

It has been recognized that there should be some sort of accreditation of national certifying bodies; however, the mechanism for such accreditation is not obvious. Within Europe, of course, the national certifying bodies will be accredited to a CEN standard.

The need for centrally controlled practical tests at Level 1 has been a point of discussion, with the majority of the subcommittee members agreeing that it was a requirement.

Recertification with examination after a ten-year period is an unpopular condition, particularly in those countries with existing programs.

Operating with an ISO Standard

The major achievement represented by the publication of this ISO standard will be the agreement upon a minimum standard which will allow comparison between the certificate holders in two or more jurisdictions. Specifying authorities, purchasers and regulators will be called upon to exercise judgement on equivalence of certification. Auditing of new certification agencies will be required until confidence is established. Supplementary examinations or conditions may need to be imposed.

On the other hand, trade between countries will be facilitated because locally qualified NDT personnel will be available; purchasers will develop a greater degree of confidence in the levels of inspection carried out at the point of production; and inspectors operating in or exporting to more than one jurisdiction will be relieved of the need for duplicate certification.

The Challenges Ahead

The publication of an international standard for the qualification and certification of NDT personnel will be a major pace forward, but it is only the first step to harmonization of national programs.

National certifying bodies will need to set conditions for recognizing equivalence by waiving conditions or examinations for certificate holders from other specified jurisdictions, for example. Codes, standards, and regulations which reference personnel qualification documents will need to address the issue of equivalence. Accreditation and auditing schemes will need to be applied to national certifying bodies, if they are to be credible. National certifying bodies, particularly in countries where there is no experience, will need assistance in establishing valid certification systems.

References

[1] "Nondestructive Testing—Qualification and Certification of Personnel," *Draft International Standard ISO/DIS 9712-2*, International Organization for Standardization, 1991.

[2] McMaster, R. C., "Qualification and Certification of NDT Personnel," *Proceedings*, Fifth International Conference on NDT, Montreal, Canada, 1967, Queen's Printer, Ottawa, 1969, pp. 478–479.

[3] Zirnhelt, J. H. and Beswick, C. K., "International Trends in the Training, Qualification and Certification of NDT Personnel," *Proceedings*, Ninth International Conference on NDT in the Nuclear Industry, Tokyo, 1988.

[4] Havercroft, W., "Report of Meetings of the International Committee, *Proceedings*, Fifth International Conference on NDT, Montreal, Canada, 1967, Queen's Printer, Ottawa, 1969, p. 479.

[5] Resnick, I., "International Nondestructive Testing Standards," *Nondestructive Testing Standards— A Review, STP 624*, H. Berger, Ed., American Society for Testing and Materials, Philadelphia, 1977.

[6] *The Complete Recommendations on International Harmonization of Training, Qualification and Certification of NDT Personnel*, International Committee on Non-Destructive Testing, Amsterdam, 1987.

[7] *Training Guidelines in Non-Destructive Testing Techniques*, IAEA-TECDOC-407, International Atomic Energy Agency, Vienna, 1987.

[8] *Report of Voting on ISO/DIS 9712*, Standards Council of Canada, 1990-07-19.

[9] Zirnhelt, J. H., Beswick, C. K., and Wilstaetter-Greig, P., "The Commitment of Multilateral Assistance Organizations to NDT Development," *Proceedings*, Twelfth World Conference on Non-Destructive Testing, J. Boogaard and G. M. van Dijk, Eds., Elsevier Publishers BV, Amsterdam, 1989.

[10] "Meeting Report on NDT and the Single Market," *British Journal on NDT*, Vol. 32, No. 12, Dec 1990, p. 641.

NDT Standards: Advanced Applications

Harold Berger[1] and Timothy Hsieh[1]

Standards for Storage, Retrieval, and Exchange of NDT Data

REFERENCE: Berger, H. and Hsieh, T., **"Standards for Storage, Retrieval, and Exchange of NDT Data,"** *Nondestructive Testing Standards—Present and Future, ASTM STP 1151*, H. Berger and L. Mordfin, Eds., American Society for Testing and Materials, Philadelphia, 1992, pp. 203–210.

ABSTRACT: Users of nondestructive testing need to have the capability to call up easily the results of previous and other method inspections in order to better interpret NDT data. The new breed of computerized NDT equipment, such as ultrasonic scanners and X-radioscopic systems, already makes use of computerized storage of NDT data. Needed are standards to facilitate the transfer of such data so that results can be made available from different and often incompatible equipment and methods. Related standards are reviewed, and progress toward standards for the transfer of NDT data are discussed. An NDT transfer effort is in progress. This will make use of a neutral format exchange as envisioned in the work of the international effort to develop the exchange document Standard for the Exchange of Product Model Data (STEP), the specification planned to include manufacturing and life cycle data.

KEY WORDS: nondestructive testing data, data storage, data retrieval, data transfer, intermediate exchange

Nondestructive testing (NDT) records of inspection are often required to be maintained for many years after an object is manufactured. These inspection records have typically involved the storage of radiographic film, ultrasonic C scans on paper, inspector notebooks, etc. These hard copy inspection records are bulky and difficult to store and locate when needed. In addition, users must be alert to the possible deterioration of such records under storage and to the accuracy and completeness of any reproductions (such as microfilm, for example) that have been made.

The standards presently available for NDT records mainly address radiographic film and the procedures that need to be followed to achieve archival records. Examples include Specifications for Photographic Film for Archival Records, Silver-Gelatin Type on Cellular Ester Base (ANSI PH1.28), and the companion standard for film on polyester base (ANSI PH1.41). These standards have served the NDT industry and helped meet the need for long-time (40-year) storage of radiographic film records. Related ASTM standards for industrial radiography include: Guide for Radiographic Testing (E 94), Guide for Controlling the Quality of Industrial Radiographic Film Processing (E 999), and Guide for the Storage of Radiographs and Unexposed Industrial Radiographic Films (E 1254).

At present it is recognized that the rapid increase in the use of computerized inspection equipment now makes it feasible to maintain NDT results as digital records on magnetic or optical media. This new capability to call up previous inspection data obtained in manufacture

[1] President and materials scientist, respectively, Industrial Quality, Inc., 19634 Club House Rd., Gaithersburg, MD 20879. (Mr. Hsieh's present address is: FMC Ground Systems Division, P.O. Box 58123, MD 570, Santa Clara, CA 95052.)

or in a previous maintenance inspection offers great promise for improvement in interpretation and reliability of inspection. Computerized records will make it possible for an inspector to access records from different types of testing records (X-ray, ultrasound, eddy current, infrared, etc.) of the same object. The information gained from using multiple NDT methods and comparing them will lead to enhanced interpretation capability. ASTM and other standards organizations have recognized these changes and are beginning to address the standards needs. These new approaches to the problem of storage and transfer of NDT data will be emphasized in this article.

Data Transfer

There has been a strong effort in electronic data transfer, largely supported by the Department of Defense (DoD) Computer Aided Acquisition and Logistical Support projects (CALS) [1,2]. Another major effort in data exchange concerns the transfer of CAD/CAM data through a neutral format specification. This system requires a translator to move the data from a given (probably proprietary) system to and from a neutral format exchange. The intermediate approach offers several advantages over a direct transfer between two or more systems. These include: (1) easy entry into the system by new manufacturers or models—only two translators will be needed, (2) protection of proprietary information (since actual exchange will be through an intermediate system), and (3) in the case of the specification in use for CAD/CAM and related data, compatibility with standards used or planned by ANSI, ISO, and MIL standards. Figure 1 demonstrates the principle of this exchange approach. Translators or processors permit data to be moved to and from a proprietary data handling system and the neutral intermediate exchange.

The consensus specification now being used for the exchange of CAD/CAM data is the Initial Graphics Exchange Specification (IGES). This specification has evolved over the last decade [3,4]. IGES, now an ANSI standard, allows exchange of information among computer-aided design systems [5]. It defines file structure format, a language format, and the represen-

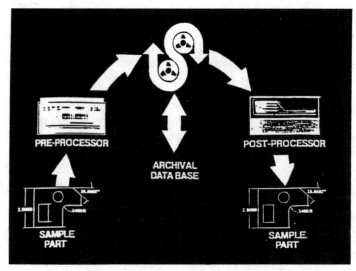

FIG. 1—*Illustration of the intermediate exchange approach.*

tation of geometric, topical, and nongeometrical product definition data in these formats to describe and communicate engineering characteristics of physical objects as manufactured products. It allows incompatible CAD systems from different vendors to translate data through a neutral format using pre- and post-processors. An important feature of the intermediate exchange approach is that proprietary interests of equipment manufacturers are protected since only the design of the processor to their equipment requires detailed knowledge of proprietary data handling approaches.

Under development and intended to take over many of the functions of IGES is the Standard for the Exchange of Product Model Data (STEP). This is being developed as an international standard whose goal is to develop a neutral mechanism to completely represent product data throughout its life cycle. PDES (Product Data Exchange using STEP) is the development activity in the United States in support of STEP. In 1990 the acronym was altered from Product Data Exchange Specification (PDES) [6] to its current one. This was done to clarify its intent of supporting the development and implementation of STEP in the United States. The goals of the PDES organization are to ensure that the requirements of U.S. industry are incorporated into the standard and to provide a methodology for U.S. industry to implement STEP standards.

The detail of the IGES/PDES Organization (IPO) is shown in Fig. 2 [7]. The three officers and five standing committees that make up the Steering Committee manage the IPO by setting policies and approving its procedures, personnel assignments, goals, and milestones. The General Assembly is composed of technical committees and three interest groups. It is overseen by a chair, three project managers, and a chair for each technical committee and interest group.

Additional work on the storage, retrieval, and transfer of data has been led by ASTM Committee E49, Computerization of Material Property Data. As the name implies, Committee E49 has worked to assist the technical community in developing standards for creating and accessing computerized material property data bases [8]. Committee E49 has also taken on for itself the assignment to assist ASTM committees and other organizations in their efforts to develop computer-oriented standards. The committee has actively assisted ASTM Committee

IGES/PDES Organization

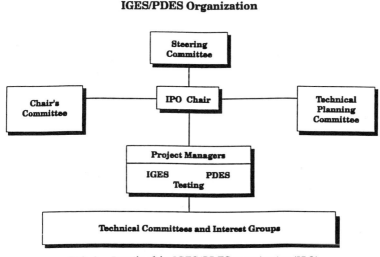

FIG. 2—*Details of the IGES/PDES organization (IPO).*

E7, Nondestructive Testing, in their initial efforts to develop storage, retrieval, and transfer standards for NDT data [9].

Standards Related to NDT

The medical diagnostics community has similar objectives to those of the NDT community in that both groups want to be able to call up diagnostic records taken at different times and with different equipment. The medical community has been working to accomplish this goal through standards being developed by the American College of Radiology (ACR) and the National Electrical Manufacturers Association (NEMA). One result of this effort is the ACR/NEMA standard, Digital Imaging and Communications [10].

The ACR/NEMA standard provides a method for transferring images and associated medical information from a variety of diagnostic equipment. The goals of this standard are to provide the following: (1) central location for data, (2) accessibility to all clinical information for a patient, (3) transfer of image and data without corruption, (4) availability of information at many locations simultaneously, (5) easy addition of another piece of equipment to communicating devices. This standard defines a minimum number of attributes for an imaging device interface to communicate with other devices and is not intended to be an overall picture and archiving and communications standard (PACS) even though it relates strongly to that objective [11]. The transfer of images using the ACR/NEMA format is a one-way transfer between the diagnostic device and a central archiving/storage device. Once the data are transferred to the archiving/storage center, it can then be accessed by any workstation connected to the PACS network. A difference in the ACR/NEMA approach is that there is no need for an intermediate file format, since the images will be viewed on workstations.

The standard itself outlines hardware requirements (the connector, pinout assignment, and timing information), the layered transfer interface (modeled after the ISO-OSI format), and the data format. The images, text, and overlays compose a "message." Each message contains one image with the associated patient and clinical information. This information is organized into 24 groups; this is further divided into data elements. Data elements contain four fields: group number, data element, length, and value. The image geometry is specified to be rectangular with the image coordinate system defined by pixels. Two other coordinate systems are defined in the ACR/NEMA standard. The image orientation is specified with respect to the manufacturer-defined equipment coordinate system. In addition, patient orientation is specified.

Another closely related data exchange standard is the Dimensional Measurement Interface Specification (DMIS) [12]. This standard is designed for communication between a computer and automated surface measurement inspection equipment. DMIS is a syntax that allows tolerance, geometry, coordinate systems, and motion information to be exchanged between CAD/CAM systems and measurement equipment. DMIS became an official project of Computer Aided Manufacturing-International (CAM-I) in 1982 and was released to the public in 1987. It was approved as an ANSI standard in February 1990 [13].

As for data exchange standards within the NDT community, one early effort (1982) was promoted under the auspices of the Electric Power Research Institute (EPRI) for the storage and transfer of ultrasonic data [14]. EPRI came to the conclusion that some of the advanced ultrasonic testing (UT) systems that EPRI had developed employed signal processing techniques which required that the UT signal be digitized. Since the ultrasonic signals were digitized before processing, the signals were in a form which could easily be stored as permanent record. EPRI recognized that the archival records would provide a source of data (1) to compare subsequent nondestructive and destructive tests, (2) to develop and evaluate new signal processing techniques, (3) for statistical analysis, and (4) to provide a means to exchange data between researchers at different locations.

The EPRI specifications for recording medium and format used nine-track magnetic tapes with the data to be stored in ASCII records of 80 characters. No standard block size was designated, but sizes between 1280 to 8192 were acceptable with the size to be placed on the tape's label. The information on each record included a header, which consisted of experimental and ultrasonic signal information, the raw data, and any reference or calibration signals. The header for the experiment contained details about the transducers, pulser, receiver, digitizer, and specimen data. The ultrasonic signal header included information on the unique waveform number, positioning information, system gain, and system delay. Since this specification was intended only for pipe and plate inspection, only the coordinates of the transducers relative to zero reference position (ZRP) stamped on the specimen were recorded. This specification, entitled Digital Recording of Ultrasonic Signals (DRUS), represented a significant early start for the computerized storage and transfer of ultrasonic NDT data.

Computerized records have been stored primarily on magnetic tape and disks. Standards for these computer-oriented media are addressed in several ANSI standards. Examples include: ANSI X3.40-1976, Unrecorded Magnetic Tape; ANSI X3.39-1986, Recorded Magnetic Tape; and ANSI X3.125-1985, Two-Sided, Double Density Disk. The NDT community is also beginning to address storage media issues. A document presently under consideration in ASTM Committee E7 is entitled, Standard Guide for the Storage of the Media that Contain Analog or Digital Radioscopic Data.[2]

There are several other documents now in preparation in ASTM Committee E7 that relate directly to the storage, retrieval, and transfer of NDT data. One, involving data storage, was cited above. Other documents address needs for the data fields for NDT examination records. One addresses ultrasonic data fields[3] and the other, radiological examination data fields.[4] All these documents are in advanced stages of preparation, but none has been issued as of the date of this symposium.

There is also a currently active Navy-sponsored program to demonstrate the capability to transfer ultrasonic inspection data between several different systems. This project, under contract to Industrial Quality, Inc. (IQI) [15], is directed toward the use of a neutral format exchange specification. The particular exchange specification selected is STEP, one that is now in an intense period of preparation with strong international participation as discussed earlier. The intermediate exchange approach was determined to be appropriate for the exchange of NDT data for the same reasons cited earlier for the use of IGES for CAD/CAM data. The selection of STEP was made because this specification is intended to include product manufacturing and life cycle data, a logical home for NDT data. No formal proposal has been made to the ISO STEP Committee for the consideration of NDT data exchange. However, informal input has been made and a formal presentation is planned.

In the current IQI-Navy ultrasonic data exchange program, work is proceeding to develop the information necessary to conduct an effective data transfer in a manner compatible with STEP. The file format and contents for the intermediate data file standard have been established. A software package has been developed to assist entry into the system by new manufacturers or models. This software package insures communication with the intermediate file format. This software consists of the Logical Intermediate File Interface and the Intermediate

---

[2] Now available as ASTM Standard E 1454-92.
[3] ASTM Guide for Data Fields for Computerized Transfer of Digital Ultrasonic Testing Data (E 1454-92).
[4] ASTM Guide for Data Fields for Computerized Transfer of Digital Radiological Test Data (E 1475).

File Services. The Logical Intermediate File Interface provides a logical systematic method of getting data into and out of the intermediate data file. It consists of a set of modules that perform the following functions: (1) read and write information into the intermediate file, (2) error reporting and logging, and (3) supplementary and configuration functions. Additional modules provide a custom programming interface for a file regardless of its physical file format and a STEP specific I/O module.

The Intermediate File Services provide all I/O and low-level formatting to the STEP format intermediate data file. The vendor designed translator is software that is vendor specific. It is designed to use the Logical Intermediate File Interface to read and write an intermediate data file. The vendor proprietary data base services is a software package that vendors should have for reading and writing their proprietary ultrasonic data file formats. The flow of information to and from the intermediate specification and the vendor ultrasonic data is illustrated in Fig. 3.

The ultrasonic data field procedures are modeled after the guide currently in process in ASTM. This is part of the attempt to make use of existing standards wherever possible. The in-preparation guide describes information needed for reporting ultrasonic test results as indicated in Table 1.

Concluding Remarks

Existing and in-process standards relating to the storage, retrieval, and exchange of NDT data have been discussed. It is clear, in this age of growing awareness for improved quality, that there will be increasing need for use of all the capability of NDT systems. Our present systems will have to be adapted to make it easier to call up previous inspection records and to examine products with multiple NDT approaches. These objectives can be met by implementing sys-

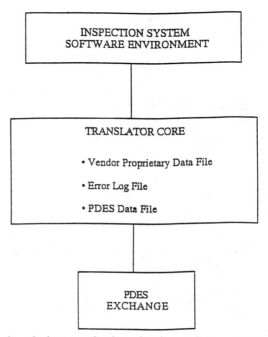

FIG. 3—*Flow of information for planned exchange of ultrasonic inspection data.*

TABLE 1—*Outline of information needed for reporting ultrasonic test results.*

---

1. Header information.
2. Inspection system description.
3. Pulser description.
4. Receiver description.
5. Gate description.
6. Search unit description.
7. Test sample description.
8. Coordinate system and scan description.
9. Test parameters.
10. Test results.

---

tems for the efficient storage, retrieval, and transfer of NDT data. Ideally, the standards and specifications for NDT data exchange should be closely tied to similar documents relating to products, manufacturing, and life cycle. It is our opinion that this can best be done through the in-process international exchange specification STEP since this is the specification planned to accommodate all product and life cycle data. Procedures to facilitate the transition to STEP are now in preparation.

The importance of the planned use of STEP was described in a recent article [*16*] and pinpointed by that article's quotation from a December 1990 speech by Department of Commerce Under Secretary for Technology, Robert M. White, as follows: "As a common standard to which all design and manufacturing software can adhere, STEP will enable users with different computers to contribute, to access, and to share mechanical, electrical and structural information not previously available in a standard format . . . The interchangeability of digital information conveys the same advantages as the interchangeability of parts."

We believe it will be vital that NDT results be part of this exchange process. It is encouraging to see significant efforts in that direction and strong related work to facilitate the exchange of NDT data.

References

[*1*] "Computer Aided Acquisition and Logistics Support Program Implementation Guide," MIL-HDBK-59, Department of Defense, Washington, DC, December 1988.

[*2*] "Automated Interchange of Technical Information," MIL-STD-1840, Department of Defense, Washington, DC, September 1986.

[*3*] "Initial Graphics Exchange Specification (IGES), Version 4.0," Report NBSIR88–3813, National Bureau of Standards, Washington, DC, 1988.

[*4*] Conroy, W., Harrod, D., and Reed, K., "Initial Graphics Exchange Specification (IGES) Version 5.0," Report NISTIR 4412, National Institute of Standards and Technology, Gaithersburg, MD, September 1990.

[*5*] "Digital Representation for Communication of Product Definition Data," ASME/ANSI Y14.26M-1989, ASME or ANSI, New York, 1989.

[*6*] Smith, B. S. and Rinaudot, G., "Product Data Exchange Specification (PDES), First Working Draft," Report NISTIR-88-4004, December 1990.

[*7*] *IGES/PDES Organization Reference Manual*, National Computer Graphics Association, Fairfax, VA, October 1990.

[*8*] Rumble, J., Jr., "Making Materials Database Standards International," *ASTM Standardization News*, Vol. 17, No. 6, June 1989, pp. 32–36.

[*9*] Rumble, J., Jr., "Presentation on Computerized Data to the Executive Committee of ASTM E7," St. Louis, MO, June 24, 1989, National Institute of Standards and Technology, Gaithersburg, MD.

[*10*] *Digital Imaging and Communications*, ACR/NEMA Standard Publication No. 300, National Electrical Manufacturers Association, Washington, DC 1988.

[*11*] *PACS—Picture Archiving and Communications System—A NEMA Primer*, NEMA, Washington, DC, November, 1988.

[*12*] Neshta, B., "The Road to a DMIS Standard," *Quality Magazine*, December 1989, pp. Q22–23.

[*13*] "Dimensional Measurement Interface Specification," ANSI/CAM-I 101, Computer Aided Manufacturing International, Inc., Arlington, TX, 1990.

[*14*] Doctor, S. R., Avioli, M. D., Barron, R. L., and Beverly, R. L., "Improving Ultrasonic Inspection Reliability," Report EPRI NP-2568, Electric Power Research Institute, Palo Alto, CA, August 1982.

[*15*] Berger, H., Hsieh, T., and Jones, T. S., "Transfer of Ultrasonic Inspection Data Among Different Systems," *Proceedings*, DOD Conference on NDT, Modesto, CA, 5–9 Nov. 1990, Sharp Army Depot, Lathrop, CA.

[*16*] Jenks, A., "STEP Forward for Joint CALS Initiative," *Washington Technology*, Vol. 5, No. 20, 24 Jan. 1991, p. 19.

Alex Vary[1]

NDE Standards for High-Temperature Materials

REFERENCE: Vary, A., "**NDE Standards for High-Temperature Materials,**" *Nondestructive Testing Standards—Present and Future, ASTM STP 1151,* H. Berger and L. Mordfin, Eds., American Society for Testing and Materials, Philadelphia, 1992, pp. 211–224.

ABSTRACT: High-temperature materials include monolithic ceramics for automotive gas turbine engines and also metallic/intermetallic and ceramic matrix composites for a range of aerospace applications. These are materials that can withstand extreme operating temperatures that will prevail in advanced high-efficiency gas turbine engines. High-temperature engine components are very likely to consist of complex composite structures with three-dimensionally interwoven and various intermixed ceramic fibers. The thermomechanical properties of components made of these materials are actually created in-place during processing and fabrication stages. The complex nature of these new materials creates strong incentives for exact standards for unambiguous evaluations of defects and microstructural characteristics. NDE techniques and standards that will ultimately be applicable to production and quality control of high-temperature materials and structures are still emerging. The needs range from flaw detection to below 100-μm levels in monolithic ceramics to global imaging of fiber architecture and matrix densification anomalies in composites. The needs are different depending on the processing stage, fabrication method, and nature of the finished product. This report discusses the standards that must be developed in concert with advances in NDE technology, materials processing research, and fabrication development. High-temperature materials and structures that fail to meet stringent specifications and standards are unlikely to compete successfully either technologically or in international markets.

KEY WORDS: NDT, NDE, ceramics, refractory composites, materials characterization, signal analysis, turbine engines, R&QA, standards

There is a need for a new generation of structural materials suitable for high-performance, high-temperature heat engines. The materials must exhibit sufficient strength, toughness, and durability to resist mechanical damage and thermal degradation while operating at extreme temperatures, i.e., at maxima of approximately 1300 or 1600°C depending, respectively, on whether metallic or ceramic materials are used [1]. In addition, the materials must permit being formed into light-weight, efficient heat engine components. These requirements can be met by toughened monolithic ceramics and by ceramic fiber-reinforced refractory composites with ceramic, metallic, and intermetallic matrixes [1,2]. Monolithic silicon carbide and silicon nitride are leading candidates for hot section components in terrestrial automotive heat engines [3]. Ceramic fiber-reinforced composites with ceramic, metallic, and intermetallic matrixes are contemplated for aerospace power and propulsion applications and associated high-temperature structures.

It has been estimated that quality control and inspection of advanced composites may represent as much as 35% of the cost of manufacture [4]. This suggests the degree of thoroughness

---

[1] Branch manager, NASA Lewis Research Center, Structural Integrity Branch, Cleveland, OH 44135.

and sophistication of inspection technology that will be necessary not only for final products but also for monitoring and controlling incoming raw materials and for component processing and fabrication. New refractory materials will tax the capabilities of current nondestructive evaluation (NDE) and inspection technology [5]. Some totally new NDE and standardization approaches will be needed. Existing inspection techniques and standards will require augmentation.

The situation demands that present inspection standards be upgraded and that new standards be developed in concert with the advancement and development of inspection methods. This should be done concurrently with the evolution of processing and fabrication methods for the new generation of high-temperature materials and structural components. Appropriate inspection standards should be considered for use (a) during raw material processing to ensure purity and quality, (b) during component fabrication to screen out defective pieces, and (c) during service to assess mechanical damage and thermochemical degradation [6,7].

Without suitable inspection methods and standards, the quality, integrity, reliability, and serviceability of new high-temperature structures will remain uncertain. High-temperature materials and structures that fail to meet internationally developed and accepted inspection standards are unlikely to successfully compete in high-technology markets [8]. This report reviews prevailing needs and recommends approaches and activities required to ensure that appropriate and necessary inspection methods and standards are developed.

Situation Assessment

Advanced structural ceramics and refractory composites for space power and propulsion systems present inspection difficulties that exceed those encountered with conventional engineering materials. Nondestructive evaluation methods and standards that will ultimately be applicable to production and quality control of the new refractory materials and structures are still emerging. The problems being addressed range from flaw detection well below 100-μm levels in monolithic ceramics to global imaging of fiber architecture and matrix anomalies in composites. The inspection needs are different depending on the processing stage, the fabrication method, and the nature of the finished product. For example, specific methods are needed for inspecting powders and green compacts before monolithic ceramics are densified by hot pressing or sintering.

For fully densified monolithic ceramic components, inspection techniques must certainly detect and characterize various types of discrete defects like cracks, voids, and other overt discontinuities. It is also important to discern and characterize microstructural conditions and diffuse flaws that govern overall strength, fracture toughness, impact resistance, and resistance to thermal-mechanical-chemical degradation [9,10]. Dispersed microflaws and morphological anomalies can reduce reliability and service life just as much as individual macroflaws. McCauley [11] has pointed out that "hidden 'defects' like subtle differences in porosity, phase composition, microstructure-retained strain (residual stress), and subcritical cracks can result in properties well below acceptable levels, even though traditional nondestructive interrogation reveals no gross inhomogeneities, cracks, or voids."

Although monolithic ceramics have fairly good high-temperature strength and superior oxidation resistance, their brittle nature and sensitivity to minute defects lead to wide variations in mechanical properties and a low fracture toughness [12]. The fracture toughness of monolithic ceramics can be improved by transformation or whisker toughening. Further improvements in strength, toughness, and durability can be achieved by ceramic fiber reinforcement in ceramic and intermetallic matrix composites. While the strength of monolithic ceramics is governed by the size and population density and distribution of minute defects, fiber-reinforced composites are insensitive to minor matrix flaws [13]. Strength, toughness, and fracture resistance of composites depend primarily on intrinsic fiber strength, fiber-matrix bond strength, and ability of the matrix to absorb fracture energy via microcracking [14].

Advanced heat engine components are likely to consist of fiber-strengthened composite structures. Strengthening will include reinforcement with a variety of intermixed ceramic fibers that are three-dimensionally interwoven. The extrinsic thermomechanical properties of these composite structures are literally created in-place during processing and fabrication stages [15]. Their complex nature creates the need for new approaches and standards that allow unambiguous evaluations of defect states, internal structural anomalies, and subtle morphological factors that govern their mechanical and load response properties.

Because high-temperature materials are still under study and development, they are moving targets for materials characterization and inspection technology. This situation calls for parallel development of nondestructive evaluation technology alongside processing and fabrication research advancements. By parallel development it becomes possible to assure that inspection methods and standards mature simultaneously with advancements in refractory materials.

Approaches to Standards Development

ASTM Activities

The formulation of reference and calibration standards for inspecting ceramics and refractory composites was formally initiated during 1988 by ASTM Committee C-28 on Advanced Ceramics. A task force for devising ceramic NDE standards was formed. The task force began by surveying all pertinent extant documents with the idea of modifying them if necessary to cover advanced ceramics. Since early 1990, over 25 ASTM Committee E-07 on Nondestructive Testing standards were reviewed and changes recommended to make them applicable to advanced ceramics and refractory composites. The recommendations were forwarded to cognizant subcommittees and are in various stages of becoming incorporated into appropriate ASTM documents. One result of the previously mentioned ASTM activities is a proposed new document entitled "Test Methods and Standards for Nondestructive Testing of Advanced Ceramics." The purpose of the document is to serve as a standard guide that identifies radiological, ultrasonic, and liquid penetrant inspection methods and procedures for advanced ceramics and refractory composites. The guide identifies current ASTM standards that are directly applicable to the examination of ceramics and refractory composites. The guide also covers ASTM standards that have been modified by mutual agreement between Committees E-07 and C-28.

A second result of ASTM activities is the development of a new document entitled "Fabricating Ceramic Reference Specimens Containing Seeded Voids." This document provides an ASTM standard practice for fabricating green and sintered bars of silicon carbide and silicon nitride containing internal and surface-connected voids at prescribed locations. The test bars will contain intentionally introduced discontinuities with known sizes and shapes. The purpose is to provide calibration standards for determining the relative detection sensitivity and spatial resolution of ultrasonic and radiographic techniques. Bars of this type have been used to establish probability-of-detection statistics and inspection parameters and procedures for a range of material conditions in monolithic ceramics [16,17].

A third result of ASTM activities is a new tabulation of densities and ultrasonic velocities for advanced ceramics and high-temperature composites. These are essential engineering data that are currently unavailable in ASTM Practice for Measuring Ultrasonic Velocity in Materials (E 494). This is a continuing effort to ensure that accurate, comprehensive density and velocity data are available for a broad range of ceramics and refractory composites.

Unique sets of characteristic ultrasonic velocities are exhibited by fully dense, monolithic materials, e.g., polycrystalline metals and glasses. However, ceramics that have porosity and fiber-reinforced composites that have both texture and porosity will exhibit a range of velocities according to the degree of porosity and anisotropy [18,19]. Examination and evaluation of ceramics and refractory composites are ultimately dependent on compilations of data connecting velocity with texture and porosity. These data are needed because of the interrelations among velocity, texture, density, elastic moduli, and mechanical properties.

Current ASTM activities will certainly help assure that needed inspection techniques and standards are established for high-temperature materials. In some instances it appears that modifications of existing documents will suffice. These modifications are necessary but insufficient because the documents were originally developed for conventional materials and methods. The proper inspection of advanced materials and structures will require some totally new standards based on innovative NDE methods.

Monolithic and Toughened Ceramics

For monolithic ceramics the chief problem is to detect distinct flaws such as cracks, voids, grain clusters, and foreign inclusions having sizes to 100-μm levels and often down to 10-μm levels [20]. Appropriate flaw detection methods are needed to deal with surface, subsurface, and volume flaws. Dispersed microporosity, diffuse flaw populations, texture and density variations also need to be found for their potentially deleterious effects on the strength and fracture resistance of monolithic and toughened ceramics.

Among the most important requirements for the specification of inspection methods for ceramics is the establishment of probability-of-detection (POD) data for a variety of flaw types. Probability-of-detection data must be accompanied by a description of exact material conditions (surface finish, thickness, shape, grain structure/coarseness, etc.) under which they were determined for specific inspection procedures and instrument settings. Only with this approach can a basis be established for selecting appropriate inspection parameters and for determining their potential effectiveness.

Fractography conducted on monolithic ceramic bend specimens has shown that principal fracture origins were subsurface and surface pores or voids [21]. These were followed, in approximate order of frequency, by narrow crack-like voids, columnar grains, large grains, clusters of grains, metallic inclusions, and surface-machining scars. The principal fracture origins just mentioned are common to the MOR (modulus of rupture) bars used, i.e., bend specimens that are sensitive to surface flaws. Volume flaws rather than machining scars and superficial flaws would dominate in other cases, depending on stress patterns. For each type of potential fracture origin, surface or volume, external or internal, it is necessary to establish POD statistics for each individual inspection technique.

In addition to detecting dominant individual flaws such as inclusions, voids, and cracks, it is essential to characterize monolithic ceramics relative to dispersed porosity patterns, density gradients, and grain-size fluctuations. These latter factors form the environments of discrete flaws and govern susceptibility to crack growth and fracture. In the case of particulate, transformation, and whisker-toughened ceramics, it is necessary to detect and characterize microstructural anomalies, density variations, adverse textures, and anomalous whisker alignments [22].

Appropriate nondestructive evaluation techniques are required to quantitatively characterize the above-mentioned microstructural and morphological features in monolithic and toughened ceramics. These techniques should provide imaging and mapping methods that reveal global variations of porosity, texture, and diffuse flaw populations. The imaging need not resolve each individual microflaw in diffuse populations. In this case resolution of the individual microflaws is usually impractical and unnecessary. Instead, what is needed is a quantitative assessment of the extent and distribution of these aberrations. This materials characterization approach is useful for comparing parts before they are placed in service and assessing changes due to thermomechanical degradation from exposure to service environments.

Ceramic and Intermetallic Matrix Composites

Composites must be inspected for constituent integrity, delaminations, disbonds, and other overt discontinuities as well as for harmful local and global variations in matrix densification, fiber distribution, fiber architecture, intralaminar integrity, and fiber-matrix bond quality [23]. It is relatively easy to create artificial disbonds in composite laminates by inserting foreign materials having various sizes and shapes, e.g., plastic wafers, metal foils, or debonding agents. These are contrivances used to simulate real discontinuities in calibration samples. They have been used as means for establishing detectability data, instrument settings, and inspection parameters.

Composites can be approached with the attitude that the detection of individual microflaws is unnecessary. This does not mean that distinct macroflaws such as delaminations, cracks, and similar discontinuities can be ignored. It should simply be recognized that composites may contain a profusion of minute defects that have no discernable effect on reliability or performance unless they are in close proximity and interact massively or permit environmental degradation at high temperatures.

What must be detected in composites are associations of flaws that can collectively degrade reliability and performance. Sparsely distributed, occasional matrix cracks, broken fibers, or misaligned fibers need be of little concern. Improper bonding between fiber and matrix must be of high concern [13]. In ceramic matrix composites, the fiber-matrix bond should be neither too strong nor too weak, while in metallic and intermetallic matrix composites the bond may be quite strong. Generally, a key factor is the quality of fiber-matrix interfaces and interphases that, in turn, determine overall strength, fracture toughness, and impact resistance. In refractory composites, therefore, a major challenge is to characterize the collective effect of improper fiber-matrix and interlaminar bonds on the mechanical integrity and strength. This is in addition to the need to detect any overt, dominant discontinuities or global aberrations that would have an overriding effect on structural integrity under particular loading conditions.

Probably the greatest challenge to the inspection of composites is the difficulty of generating reference and calibration standards that possess subtle microstructural aberrations that nevertheless can have significant effects on mechanical properties and load response, e.g., fiber-matrix interface bond irregularities. The calibration standards should be in the form of material samples that possess representative structural aberrations and corresponding mechanical property variations while duplicating the anisotropies and geometric properties of real parts.

Materials Characterization

McCauley [11] has argued that advanced refractory materials represent enormous challenges and that it is necessary to "recognize the importance of materials characterization concepts for controlling and monitoring a material's full unique signature" and that "this will require the extension of traditional NDT into chemical and microstructural interrogation, transitioning sophisticated materials characterization techniques out of the research laboratory."

It is difficult enough even in the case of monolithic, polycrystalline solids, e.g., metals, ceramics, to generate reference standards for quantitative ranking of microstructure-dependent properties (that is, strength, toughness, impact resistance). The difficulty is compounded for composites with complex, heterogeneous, anisotropic microstructures. These complications need to be overcome in developing representative materials and benchmark structures that can be used as comparative reference standards for materials characterization and instrument calibration.

A specific challenge to inspection standardization and calibration technology is the need to fabricate reference samples that exhibit microstructures and morphologies that represent a realistic range of material conditions and mechanical properties from poor to ideal. This is to ensure that nondestructive materials characterization techniques will be able to differentiate rejectable from desirable parts. Underlying this approach is the fact that nondestructive methods are indirect and depend on signal interpretations and empirical correlations to assess the quality and mechanical characteristics of a material or structure.

The simplest approach is to comparatively characterize a set of test samples that have been subjected to different levels of thermal or mechanical degradation. Each sample in the set would initially have been identical to all the others, based on careful verification by suitable NDE methods. After thermal or mechanical conditioning, each sample exhibits different physical-chemical-mechanical properties, e.g., modified fiber-matrix interface properties. Although each sample in the set constitutes an important reference, the sample with optimum properties is taken as a benchmark. This assumes that the benchmark sample is either in a pristine condition or otherwise represents an ideal, preferred condition of the material or structure.

Because the quality and strength of monolithic, composite, and composite-like material are subject to numerous processing variables, it is useful to feed back nondestructive evaluations to process development research. This concomitantly aids in creating temporary "application" standards for identifying the most successful production conditions and the best resultant materials and parts. Nondestructive monitoring during processing research and fabrication development can help identify and refine the best ultimate inspection standards and property characterization procedures.

The structural integrity of monolithic ceramics and refractory composites depends on avoiding fabrication flaws and maintaining high quality during processing [24]. An approach for consistently producing high-quality ceramics is to utilize nondestructive evaluation techniques during materials research and processing development to help determine stages when harmful flaws are likely to be introduced. Steps can then be taken to minimize their occurrence through improvements in processing. This can be done at various stages of processing to save the cost of finishing parts that contain defects from an earlier stage. The least efficient approach, usually avoidable, is to use nondestructive evaluation after the last stage of fabrication to reject parts that contain harmful flaws. This can result in costly high rejection rates because one cannot "inspect in" quality!

Signal Analysis and Evaluation

The peculiarities and complexities of advanced materials, especially composites and composite structures, will require approaches that go beyond simple calibration pieces. For advanced materials, calibration samples and elementary procedures may not suffice. Indeed, simple "universal" calibration standards can be invalid and illusory. This observation is based on the fact that many individual factors can simultaneously influence probe media used to interrogate materials for assessing their microstructural, morphological, and mechanical property variations.

In the case of computed tomography, the effects of X-ray beam hardening and geometric shadowing can undermine image reconstruction algorithms. In the case of ultrasonics, multiple reflections, mode conversions, and boundary conditions can hinder correct measurements. In either previous case, and in general, clear correlations may be obscured by a host of incidental geometric and microstructural factors. Sophisticated interpretational methods will then be required to extract from images and signals the desired information regarding particular material characteristics or properties. In addition to advanced signal analysis approaches, multiparametric probing using several nondestructive evaluation techniques may be mandatory to extract and separate complementary and corroborative data. This will help remove ambiguities that would arise if only one technique were relied upon and where the effects of several material variables overlap and need to be isolated.

Material calibration standards certainly need to be augmented with advanced multiparametric signal analysis software and computerized evaluation methods. The appropriate foundations for these advanced methods are expert systems based on adaptive learning methods and neural networks that are, in turn, based on carefully devised learning sets. The learning sets should consist of extensive series of material samples that exhibit all combinations of factors that influence probe media and factors that are likely to exist in the materials and structures to be interrogated. Nondestructive evaluation approaches evolved from this data base may very well consist of standardized signal processing and interpretation software packages. The packages would contain algorithms for signal transformation, image enhancement, signature analysis, feature extraction, pattern recognition, and classification [25,26].

NDE Technology

General

The primary nondestructive evaluation techniques applicable to ceramics and refractory composites are visual-optical examination, liquid penetrant inspection, radiography, and ultrasonics [5,6,9]. Specialized techniques include fluorescent penetrants, microfocus X-radiography, computed tomography, analytical ultrasonics, and acoustic microscopy for monolithic and particulate and whisker-toughened ceramics. Computed tomography, film and digital radiography, scanning ultrasonics, and acousto-ultrasonics are among the specialized techniques suitable for inspecting ceramic fiber-reinforced ceramic and intermetallic matrix composites.

Methods for Raw Materials

The screening and characterization of ceramic powders and ceramic-toughening agents (crystallites, whiskers) are the first step in assuring the quality of monolithic structural ceramics. Particle size and size distribution, chemical purity, crystalline phase, morphology, contaminants, and physical properties are among the attributes that require assessment and close control. Inspection methods include light scattering, gas absorption, microscopy, X-ray diffraction, Auger and mass spectroscopy, and chemical analysis [27]. These are primarily physical-chemical analysis methods that nevertheless fall under the purview of nondestructive characterization and require appropriate standards. Similar methods are needed to assess continuous ceramic fibers, fiber tows/bundles, and fiber preforms used to fabricate composites.

Additional raw materials involved in fabricating monolithic ceramic structures and refractory composites are processing aids such as organic binders, dispersants, lubricants, and also carrier vehicles such as water, solvents, vapors, and gases. These latter ingredients must be characterized for purity, contamination, molecular weight, viscosity, and their relative effectiveness during processing stages such as forming, injection molding, slip casting, infiltration, and chemical vapor deposition.

Methods for Green Compacts

The formation of bisques and green state bodies is an intermediate step in the fabrication of structural ceramics and refractory composites. The overall shapes of structural components are created at this stage, followed by sintering or hot pressing to form densified near net or final shapes. This is a crucial stage during which flaws can be introduced or substandard materials can be accidentally produced. Binder maldistribution, density fluctuations, porosity, inclusions, and similar volume discrepancies must be assessed and controlled [28]. Also, green compact dimensions and surface roughness are factors that can attest to the goodness or poorness of processing conditions and controls.

X-ray absorption and nuclear magnetic resonance are sensitive to binder distribution anomalies in green compacts [29]. Laboratory studies have shown that porosity and other volume flaws in these compacts can be detected using film and digital radiography, computed tomography, and nuclear magnetic resonance methods. Green state compacts and bisques are quite fragile so that inspections are best accomplished with techniques that avoid forceful physical contact.

Metrology methods using noncontacting laser optical techniques provide fast and sensitive means for monitoring and verifying correct green compact shapes and dimensions. Ultrasonics usually requires contact but can be accomplished without damage to green state forms under certain conditions, e.g., by use of air-coupled probes. Light scattering and laser optical techniques lend themselves to surface roughness measurements for green state and also for fully densified sintered structures. All these nondestructive evaluation methods can provide valuable feedback for perfecting processing parameters and then for monitoring various fabrication steps.

Methods for Densified Materials

Conventional, appropriate, and mandatory techniques for surface-connected flaws are optically aided visual and liquid penetrant inspections [30]. They should be used routinely to screen out articles that are cracked, pitted, marred, spalled, or have poorly finished surfaces. Immersion scan ultrasonics, film radiography, and computed tomography detect subsurface and volume flaws. If the flaws are isolated and fairly large, i.e., of the order of 500 μm or more, then conventional ultrasonic scanning and film radiographic methods are suitable. However, the spatial and image density resolution of these conventional methods becomes taxed in the "grey area" represented by flaws in the 500 to 50-μm range.

High-Resolution Flaw Detection Methods

For discrete flaws below 100 μm, it is necessary to consider high-resolution methods like acoustic microscopy, microfocus radiography, and microtomography. Acoustic microscopy and microfocus radiography can detect flaws down to the 20-μm level in monolithic silicon carbide and silicon nitride. These methods are successful and have high-resolution and high probability of detection only under the most stringent conditions of material thickness, part shape, surface finish, etc. [16,17]. Even under the best conditions and with high-resolution methods, some flaws remain very difficult to detect, e.g., tight cracks, megagrains, and grain clusters having densities or acoustic impedances that match their surroundings.

Recent laboratory results with new high-strength monolithic ceramics have shown that many failures are initiated by surface and near-surface defects between 20 and 40 μm in size. Acoustic microscopy affords the potential for detection of flaws of this nature given the right conditions. Surface preparation by polishing or fine grinding is needed to enhance the detectability even of exposed surface voids on the order of 50-μm diameter and less. Surface roughness affects the signal-to-noise ratio in acoustic microscope images. Moreover, sintered samples with as-fired surfaces show decreased volume flaw detectability with increased thickness and flaw depth. Flaw detectability also depends on the relative coarseness of the material's grain structure. In coarse-grained (silicon carbide) samples, flaw detectability was found to be significantly less than in (silicon nitride) samples that had a much finer grain structure [31].

Scanning acoustic microscopy can image flaws in monolithic ceramics with a resolution of about 20 μm or better. Scanning acoustic microscopes usually operate at 50 to 200 MHz and can be focused up to several millimeters into fine-grained monolithic ceramics. A scanning acoustic microscope operating at a center frequency of 50 MHz is readily able to image voids 20 μm in diameter at a depth of 1 mm in silicon nitride [32].

Near surface, i.e., subsurface, flaws may require examination by the ultrasonic surface wave method [33]. A focused ultrasonic transducer operating at frequencies up to 100 MHz is used to launch and collect Rayleigh waves that can interact with and resolve minute cracks and other defects down to the 10-μm level. The surface wave method overcomes difficulties encountered by the pulse reflection or scanning acoustic microscopy method primarily because the waves travel parallel rather than normal to the surface.

Microfocus radiography provides a high-resolution imaging tool with the potential of being readily applied in production as well as laboratory environments. Film and real-time video versions are available for inspecting a variety of test objects for flaws distributed throughout a volume. Recent research has shown the combined spatial and image density resolution of microfocus radiography to be at least twice that of conventional film radiography [34]. Like other projection radiographic methods, microfocus radiography is generally suitable for detecting flaws that have three-dimensional extent, e.g., voids, inclusions, as opposed to two-dimensional or planar flaws like cracks.

Computed tomography systems can produce the high-resolution images required for characterization of structural ceramics and their composites [35]. Unlike film and projection radiography, computed tomography produces cross-sectional and three-dimensional reconstructions of both discrete and diffuse flaw populations in an examined volume. High-speed computed tomographic systems readily provide image resolutions on the order of 250 μm. Using microfocus X-ray sources, advanced tomographic systems are being developed for resolving down to 25 μm [36].

Materials Characterization Methods

Fairly large flaws are frequently encountered in components such as turbine rotors. But, in monolithic ceramics, flaws less than 10 μm have been routinely found to be fracture origins. Such flaws tend to be quite numerous in fine-grained ceramics, and this situation will overburden the capabilities of any high-resolution technique. High-resolution imaging is inherently time consuming. It is important to decide whether there is sufficient payoff to examine each and every cubic millimeter of a monolithic ceramic article for each 10-μm flaw. Of course, there will be critical zones where high-resolution examination is justified.

Below the 50-μm level, it may be impractical and even unnecessary to image and characterize individual flaws in noncritical zones of monolithic ceramic structures and certainly unnecessary in refractory composite structures. The alternative is to use low-resolution methods to characterize the global environment in which flaws reside. This is the primary goal of analytical ultrasonics and macroscopic computed tomography [9,37]. These two technologies can quantitatively characterize and image diffuse flaw populations, dispersed microporosity, anisotropy, texture, sintering anomalies, fiber misalignment, etc.

The term analytical ultrasonics denotes a methodology for quantitative characterization of the microstructure and mechanical properties of engineering materials. Ultrasonic velocity and attenuation are analytical methods for assessing bulk density, grain size, and other extrinsic factors that govern strength and toughness. Models explaining and predicting the empirical correlations found between ultrasound and mechanical properties have been advanced [10,38]. These correlations depend heavily on the experimental conditions and the nature of the material sample, e.g., size, shape. Factors that influence ultrasonic attenuation and velocity measurements include surface finish, pore fraction, pore size and shape, grain size, grain-size distribution, texture, and elastic anisotropy. Of course, these same factors also govern mechanical properties, load response, and thermal and mechanical degradation.

Low ultrasonic attenuation is characteristic of nearly fully dense monolithic ceramics with fine microstructures, i.e., samples with a mean grain size of less than 10 μm and densities greater than 95% of theoretical. For monolithic and toughened ceramics, significant attenuation differences are evident only at frequencies greater than approximately 100 MHz. Fairly high frequencies are needed to correctly assess subtle microstructural aberrations such as excess detrimental granularity and porosity.

Ultrasonic attenuation is influenced by bulk density and the combined effects of pore size and grain size and, therefore, is a sensitive indicator of microstructural variations in structural ceramics when measurements are made at the appropriate frequencies [39]. However, meaningful attenuation measurements require not only fairly smooth surfaces but also constraints on sample size, shape, and thickness. When accurate attenuation measurements are needed, the surface roughness should be minimized [40]. Nevertheless, it is possible to make comparative attenuation measurements on as-fired or unpolished machined specimens provided that the surface roughness is the same on all samples and the signal-to-noise ratio is sufficiently high.

Ultrasonic velocity is a monotonically increasing function of density in porous solids [18]. Variations in pore size and grain size have little effect on this relation. Although poor surface finish and overall sample thickness can reduce accuracy somewhat, velocity measurements are not as vulnerable to surface roughness as are attenuation measurements. Since velocity measurements are not strongly affected by pore or grain size, they are convenient for estimating bulk density of monolithic and toughened ceramics. Experimental results show that velocity measurements can be used to estimate bulk density within approximately 1%. Velocity measurements can be used to screen out low-density monolithic ceramic components and refractory composite structures.

Ideally, both attenuation and velocity measurements require essentially flat, parallel opposing surfaces or geometric simplicity. Actual part shapes do not always permit precision attenuation or velocity measurements. An alternative approach is the ultrasonic backscatter method for ultrasonic determination of porosity, grain, and similar microstructural variables [41,42]. Backscattered, and under some conditions forward scattered, ultrasound radiations can be used to characterize volume properties of parts having complex shapes [43].

The acousto-ultrasonic technique was developed specifically for characterizing defect states and mechanical property variations of composites [44,45]. The acousto-ultrasonic technique has been applied to fiber-reinforced composite laminates to detect local and global anomalies such as matrix crazing and porosity, modulus or stiffness variations, interlaminar bond and fiber-matrix bond strength variations, and fatigue and impact damage. Acousto-ultrasonics is similar to coin tap, sonic vibration, and dynamic resonance methods for assessing the overall global condition of fabricated shapes [46,47]. The acousto-ultrasonic technique is a comparative analytical ultrasonic method that does not impose the stringent constraints on material surface conditions required for the attenuation measurements mentioned previously.

Conventional film radiography and projection radiography are important imaging methods for macroflaw detection, for assessing global density variations, and for locating porosity in monolithic and toughened ceramics. Digital radiography provides an excellent quantitative means for comparing degrees of densification in a volume of material. Computed tomography applied at lower resolutions can produce three-dimensional images of density variations, fiber architecture, dispersed flaw populations, and any global aberrations in refractory composite structures [48].

Auxiliary Methods

The previously mentioned nondestructive evaluation methods are prominent among the ones currently being considered and applied to high-temperature materials. This does not preclude various other methods that can be equally viable and appropriate. For example, eddy-current testing has been applied to polymer matrix composites and may prove particularly useful for characterizing intermetallic matrix composites [49]. There are numerous thermal wave techniques that already have been applied to monolithic ceramics and that may readily apply to refractory composites [50]. Electric and magnetic testing, dielectrometry, and microwave techniques have been demonstrated for polymeric composites and should also be considered for monolithic ceramics and refractory composites [51].

Acoustic emission techniques have applications ranging from materials research to component proof testing [52]. In materials research, acoustic emission can be used to monitor fracture processes and to help identify factors that govern or contribute to material failure. Acoustic emission monitoring during proof testing can aid in assessing the infirmity or integrity of high-temperature components.

Conclusion

Current activities under the leadership of ASTM committees will help assure that nondestructive evaluation and inspection standards are established for high-temperature materials. In some instances it appears that modifications of existing documents will suffice. There are other instances where new inspection methods and associated standards will be required. These depend on the development of sophisticated inspection strategies demanded by advanced ceramic and refractory composite structures. The technological needs are described in this report, and suitable approaches are suggested. The major observation is that pivotal roles will be played by advanced techniques for high-resolution flaw detection and innovative techniques for nondestructive materials characterization.

Materials characterization and high-resolution flaw detection are currently primarily laboratory techniques that require further investigation, development, and adaptation before they can be applied in materials processing, fabrication, and field environments. Practical implementation of these methods in production and field uses awaits the development of suitable calibration standards and standards of practice. Flaw detection techniques for monolithic and toughened ceramics depend on investigations that will establish statistical foundations for probability of detection of various types of defects over a range of material and component conditions. Emerging approaches for nondestructive materials characterization of ceramics and refractory composites require thorough investigation and development before they can be relied on to assess initial quality, mechanical properties, diffuse defect states, or thermomechanical degradation in high-temperature structures. Computerized interpretational procedures using expert systems will undoubtedly be needed to assure unambiguous nondestructive characterizations of specific material properties.

References

[1] *Proceedings, HITEMP Review 1990—Advanced High-temperature Engine Materials Technology Program,* NASA Conference Publication 10051, NASA, Washington, DC, 1990.

[2] *Proceedings, Department of Defense Metal Matrix Composites Technology Conference,* MMCIAC 717, Vols. 1 and 2, Defense Technical Information Center, Arlington, VA, 1989.

[3] "Research and Development of Automotive Gas Turbines in Japan," *JETRO* (special issue), Japan External Trade Organization, Tokyo, 1988.

[4] Hauwiller, P. B. "Non-Technological Impediments to Composites Producibility," *Proceedings, 22nd International SAMPE Technical Conference,* SAMPE, Covina, California, 1990, pp. 213–223.

[5] *Proceedings, Nondestructive Testing of High-Performance Ceramics,* Vary, A. and Snyder, J., Eds., American Ceramics Society, Westerville, OH, 1987.

[6] Johnson, D. R., McClung, R. W., Janney, M. A. and Hanusiak, W. M., "Needs Assessment for Nondestructive Testing and Materials Characterization for Improved Reliability in Structural Ceramics for Heat Engines," ORNL TM-10354, Oak Ridge National Laboratory, TN, 1987.

[7] Proulx, D., Roy, C., and Zimcik, D. G., "Assessment of the State of the Art of Non-Destructive Evaluation of Advanced Composite Materials," *Canadian Aeronautics and Space Journal,* Vol. 3, No. 4, 1985, pp. 325–334.

[8] *Nondestructive Testing Standards—A Review, ASTM STP 624,* Berger, H., Ed., ASTM, Philadelphia, PA, 1977.

[9] *Materials Analysis by Ultrasonics,* Vary, A., Ed., Noyes Data Corporation, Park Ridge, NJ, 1987.

[10] Vary, A., "Concepts for Interrelating Ultrasonic Attenuation, Microstructure, and Fracture Toughness in Polycrystalline Solids," *Materials Evaluation,* Vol. 46, No. 5, 1988, pp. 642–649.

[11] McCauley, J. W., "Materials Testing in the 21st Century," *Proceedings, Nondestructive Testing of High-Performance Ceramics,* A. Vary and J. Snyder, Eds., American Ceramics Society, Westerville, OH, 1987, pp. 1–18.

[12] Singh, J. P., "Effect of Flaws on the Fracture Behavior of Structural Ceramics: A Review," *Advanced Ceramic Materials,* Vol. 3, No. 1, 1988, pp. 18–27.

[13] Evans, A. G., and Marshall, D. B., "The Mechanical Behavior of Ceramic Matrix Composites," *Acta Metallurgica,* Vol. 37, No. 10, pp. 2567–2583.

[14] Gyekenyesi, J. P., "Failure Analysis of Continuous Fiber-Reinforced Ceramic Matrix Composite Laminates," *Proceedings, HITEMP Review 1988—Advanced High-temperature Engine Materials Technology Program,* NASA Conference Publication 10025, NASA, Washington, DC, 1988, pp. 145–163.

[15] Marshall, D. B., "NDE of Fiber and Whisker-Reinforced Ceramics," *Progress in Quantitative NDE,* Plenum Press, New York, 1986, pp. 60–69.

[16] Baaklini, G. Y., Kiser, J. D. and Roth, D. J., "Radiographic Detectability Limits for Seeded Voids in Sintered Silicon Carbide and Silicon Nitride," *Advanced Ceramic Materials,* Vol. 1, No. 1, 1986, pp. 43–49.

[17] Roth, D. J., Klima, S. J., Kiser, J. D. and Baaklini, G. Y., "Reliability of Void Detection in Structural Ceramics by Use of Scanning Laser Acoustic Microscopy," *Materials Evaluation,* Vol. 44, No. 6, 1986, pp. 762–769.

[18] Klima, S. J., Watson, G. K., Herbell, T. P. and Moore, T. J., "Ultrasonic Velocity for Estimating Density of Structural Ceramics," NASA TM-82765, NASA, Washington, DC, 1981.

[19] Roth, D. J., Stang, D. B., Swickard, S. M. and DeGuire, M. R., "Review and Statistical Analysis of the Ultrasonic Velocity Method for Estimating Porosity Fraction in Polycrystalline Materials," NASA TM 102501, NASA, Washington, DC, 1990.

[20] Klima, S. J. and Baaklini, G. Y., "Nondestructive Evaluation of Structural Ceramics," *SAMPE Quarterly,* Vol. 17, No. 3, 1986, pp. 13–19.

[21] Sanders, W. A. and Baaklini, G. Y., "Correlation of Processing and Sintering Variables with the Strength and Radiography of Silicon Nitride," NASA TM-87251, NASA, Washington, DC, 1986.

[22] Stang, D. B., Salem, J. A., and Generazio, E. R., "Ultrasonic Imaging of Textured Alumina," *Materials Evaluation,* Vol. 48, No. 12, 1990, pp. 1478–1482.

[23] Sela, N. and Ishai, O., "Interlaminar Fracture Toughness and Toughening of Laminated Composite Materials: A Review," *Composites,* Vol. 20, No. 5, 1989, pp. 423–435.

[24] Bhatt, R. T., "Effects of Fabrication Conditions on the Properties of SiC Fiber-reinforced Reaction-Bonded Silicon Nitride Matrix Composites (SiC/RBSN)," NASA TM 88814, NASA, Washington, DC, 1986.

[25] Chen, C. H. and Hsu, W-L, "Modern Spectral Analysis for Ultrasonic NDT," *Nondestructive Testing of High-Performance Ceramics,* A. Vary and J. Snyder, Eds., American Ceramics Society, Westerville, OH, 1987, pp. 401–407.

[26] Williams, J. H., Jr. and Lee, S. S., "Pattern Recognition Characterizations of Micromechanical and Morphological Material States via Analytical Ultrasonics," *Materials Analysis by Ultrasonics*, A. Vary, Ed., Noyes Data Corp, Park Ridge, NJ, 1987, pp. 193–206.

[27] Johnson, D. R., Janney, M. A., and McClung, R. W., "Needs Assessment for NDT and Characterization of Ceramics—Approach to Characterization Technology for Raw Materials," *Proceedings, Nondestructive Testing of High-Performance Ceramics*, A. Vary and J. Snyder, Eds., American Ceramics Society, Westerville, OH, 1987, pp. 19–32.

[28] McClung, R. W. and Johnson, D. R., "Needs Assessment for NDT and Characterization of Ceramics—Assessment of Inspection Technology for Green State and Sintered Ceramics," *Proceedings, Nondestructive Testing of High-Performance Ceramics*, A. Vary and J. Snyder, Eds., American Ceramics Society, Westerville, OH, 1987, pp. 33–51.

[29] Ackerman, J. L., Garrio, L., Ellingson, W. A., and Weylans, J. D., "The Use of NMR Imaging to Measure Porosity and Binder Distributions in Green State and Partially Sintered Ceramics," *Proceedings, Nondestructive Testing of High-Performance Ceramics*, A. Vary and J. Snyder, Eds., American Ceramics Society, Westerville, OH, 1987, pp. 88–113.

[30] Bowman, C. C. and Batchelor, B. G., "Automated Visual Inspection," *Research Techniques in Nondestructive Testing*, Vol. 8, R. S. Sharpe, Ed., Academic Press, London, 1985, pp. 361–444.

[31] Klima, S. J., "Factors that Affect Reliability of Nondestructive Detection of Flaws in Structural Ceramics," NASA TM-87348, NASA, Washington, DC, 1986.

[32] Vary, A. and Klima, S. J., "Application of Scanning Acoustic Microscopy to Advanced Structural Ceramics," NASA TM-89929, NASA, Washington, DC, 1987.

[33] Baaklini, G. Y. and Able, P. B., "Flaw Imaging and Ultrasonic Techniques for Characterizing Sintered Silicon Carbide," NASA TM-100177, NASA, Washington, DC, 1987.

[34] Baaklini, G. Y. and Roth, D. J., "Probability of Detection of Internal Voids in Structural Ceramics Using Microfocus Radiography," NASA TM-87164, NASA, Washington, DC, 1985.

[35] Gilboy, W. B. and Foster, J., "Industrial Applications of Computerized Tomography with X- and Gamma Radiation," *Research Techniques in Nondestructive Testing*, Vol. 6, R. S. Sharpe, Ed., Academic Press, London, 1982, pp. 255–287.

[36] Armistead, R. A., "CT: Quantitative 3-D Inspection," Advanced Materials & Processes, *Metals Progress*, March 1988, pp. 42–48.

[37] Ellingson, W. A., Roberts, R. A., Ackerman, J. L., Sawicka, B. D., Gronemyer, S., and Kriz, R. J., "Recent Developments in Nondestructive Evaluation for Structural Ceramics," *International Advances in Nondestructive Testing*, Vol. 13, W. J. McGonnagle, Ed., Gordon and Breach, New York, 1988, pp. 267–294.

[38] Vary, A., "Ultrasonic Measurement of Mechanical Properties," *International Advances in Nondestructive Testing*, Vol. 13, W. J. McGonnagle, Ed., Gordon and Breach, New York, 1988, pp. 1–38.

[39] Generazio, E. R., Roth, D. J. and Baaklini, G. Y., "Acoustic Imaging of Subtle Porosity Variations in Ceramics," *Materials Evaluation*, Vol. 46, No. 10, 1988, pp. 1338–1343.

[40] Generazio, E. R., "The Role of the Reflection Coefficient in Precision Measurement of Ultrasonic Attenuation," *Materials Evaluation*, Vol. 43, No. 8, 1985, pp. 995–1004.

[41] Goebbels, K., "Structure Analysis by Scattered Ultrasonic Radiation," *Research Techniques in Nondestructive Testing*, Vol. 4, R. S. Sharpe, Ed., Academic Press, London, 1980, pp. 87–157.

[42] Tittmann, B. R., Ahlberg, L. A., and Fertig, K., "Ultrasonic Characterization of Microstructure in Powder Metal Alloy," *Materials Analysis by Ultrasonics*, A. Vary, Ed., Noyes Data Corporation, Park Ridge, NJ, 1987, pp. 30–46.

[43] Roberts, R. A., "Focused Ultrasonic Backscatter Technique for Near-Surface Flaw Detection in Structural Ceramics," *Proceedings, Sixteenth Symposium on Nondestructive Evaluation*, C. D. Grey and G. A. Matzkanin, Eds., Southwest Research Institute, San Antonio, TX, 1987, pp. 224–230.

[44] *Acousto-ultrasonics—Theory and Application*, J. C. Duke, Jr., Ed., Plenum Press, New York, 1987.

[45] Vary, A., "Acousto-ultrasonics," *Non-Destructive Testing of Fiber-reinforced Plastics*, Vol. 2, J. Summerscales, Ed., Elsevier Applied Science Publishers, Essex, England, 1990, pp. 1–54.

[46] Uygur, E. M., "Nondestructive Dynamic Testing," *Research Techniques in Nondestructive Testing*, Vol. 4, R. S. Sharpe, Ed., Academic Press, London, 1980, pp. 205–244.

[47] Adams, R. D. and Cawley, P. "Vibration Techniques in Nondestructive Testing," *Research Techniques in Nondestructive Testing*, Vol. 8, R. S. Sharpe, Ed., Academic Press, London, 1985, pp. 303–360.

[48] Stoller, H. M., Crose, J. G. and Pfeifer, W. H., "Engineering Tomography: A Quantitative NDE Technique for Composite Materials," *Proceedings, American Society for Composites, 3rd Technical Conference*, Technomic Publishing, Lancaster, PA, 1988, pp. 537–547.

[49] Prakash, R. "Eddy-Current Testing," *Non-Destructive Testing of Fiber-Reinforced Plastics*, Vol. 2, J. Summerscales, Ed., Elsevier Applied Science Publishers, Essex, England, 1990, pp. 299–325.

[*50*] Murphy, J. C., Maclachlan, J. W., and Aamodt, L. C., "The Role of Thermal Wave Techniques in Materials Characterization," *International Advances in Nondestructive Testing*, Vol. 14, W. G. McGonnagle, Ed., Gordon and Breach, New York, 1989, pp. 175–218.

[*51*] *Non-Destructive Testing of Fiber-reinforced Plastics*, Vol. 2, J. Summerscales, Ed., Elsevier Applied Science Publishers, Essex, England, 1990, pp. 253–297 [Electrical and Magnetic Testing], pp. 327–360 [Dielectrometry], pp. 361–412 [Microwave Techniques].

[*52*] Scruby, C. B., "Quantitative Acoustic Emission Techniques," *Research Techniques in Nondestructive Testing*, Vol. 8, R. S. Sharpe, Ed., Academic Press, London, 1985, pp. 141–210.

William C. Plumstead[1]

Nondestructive Testing/Examination in the Construction Industry

REFERENCE: Plumstead, W. C., "**Nondestructive Testing/Examination in the Construction Industry,**" *Nondestructive Testing Standards—Present and Future, ASTM STP 1151,* H. Berger and L. Mordfin, Eds., American Society for Testing and Materials, Philadelphia, 1992, pp. 225–229.

ABSTRACT: Construction quality standards are dictated by engineering design or regulatory requirements. The type of service determines the specific nondestructive examination (NDE) requirements. Petrochemical, power, and high-pressure service are typical areas of construction that use codes and standards to specify quality requirements.

Most of the construction industry employs NDE subcontractors to perform required nondestructive examinations. Some companies use approved vendor lists for the selection of subcontractors. The evaluation of subcontractor programs and personnel is a cost to construction. Where it is not required, evaluations are often eliminated to reduce costs to construction.

A better system is needed to assure more consistent performance of nondestructive examinations. The present cost to industry resulting from substandard performance is excessive. Prequalification of NDE subcontractors programs and personnel will reduce the costs to construction that are associated with poor performance.

Increased utilization of ASTM Practice for Determining the Qualification of Nondestructive Testing Agencies (E 543) will improve the general qualifications of the nondestructive testing agencies and result in improved performance.

KEY WORDS: personnel, qualifications, construction, performance

Construction activity is a relatively short-term activity. Getting the job done with the lowest cost and fastest schedule consistent with job requirements is the name of the game. Formal quality control programs are not generally used where such a program is not required because it is perceived by many construction managers as an unnecessary, additional cost. Most contractors will only do what is required, because to do otherwise usually affects their competitive position.

When formal construction quality standards are required, they are determined by engineering design or regulatory requirements. The type of service dictates the specific nondestructive examination (NDE) requirements. Petrochemical, power generation, and high-pressure service are typical areas which use codes and standards to specify quality requirements. Formal requirements are established in these areas because of the nature of the service involved. Where failure in service may result in injury or death, quality standards are usually imposed in order to meet legal requirements or obtain insurance.

Several national organizations produce most of the codes and standards used in construc-

[1] Principal quality engineer, Fluor Daniel, Inc., 100 Fluor Daniel Drive, Greenville, SC 29607–2762.

tion. The American National Standards Institute (ANSI), American Society for Mechanical Engineers (ASME), American Society for Testing and Materials (ASTM), American Welding Society (AWS), and the American Petroleum Institute (API) have developed the majority of nondestructive examination codes and standards used in the construction industry. These groups produce requirements for their specific industry segment. They provide individually or in combination the methodology and acceptance criteria for expected workmanship standards.

All of these codes used in construction reference the American Society for Nondestructive Testing (ASNT) Recommended Practice No. SNT-TC-1A: Qualification and Certification for Nondestructive Testing Personnel. SNT-TC-1A is intended as a guideline for employers to develop their own personnel qualification and certification program for nondestructive testing personnel based on their specific needs. The basis for qualification of personnel is a combination of education, training, experience, and examinations.

The American Society for Mechanical Engineers (ASME) is well known for its pioneering of standards associated with the design and testing of pressure vessels. ASME is also the code of reference for nuclear power construction and in-service inspection since being adopted by many of the states as a requirement. ASME is divided into such specific sections as power piping, nuclear power components, nondestructive examination, and pressure vessels. Each ASME code section references the section for Nondestructive Examination (Section V) for methodology, but provides its own specific acceptance criteria.

ASME and AWS provide common standards in the United States for welder procedure and performance qualifications. Welder performance qualifications can be established by nondestructive examination of the welder test coupons. As an example, radiography performed to ASME Section V for Nondestructive Examination and evaluated to ASME Section IX for Weld and Brazing Qualification can be used to qualify a welder's ability to weld to a particular qualified welding procedure.

AWS D1.1 Structural Welding Code is used for fabrication and construction of Statically Loaded, Dynamically Loaded, and Tubular Structures. AWS D1.1 references ASTM standards for liquid penetrant and magnetic particle examination methodology and adds their specific acceptance criteria. The AWS D1.1 Structural Welding Code does require unique methodology and acceptance criteria for the radiographic and ultrasonic methods.

ANSI has developed a variety of specifications for use in specific industry segments. ANSI B31.1: Power Piping and ANSI and B31.3: Chemical Piping are two frequently used specifications in construction. They are specified because the particular applications involve potentially dangerous service. ANSI references ASME Section V for Nondestructive Examination methodology. ANSI provides the extent of examination and code specific acceptance criteria related in stringency to the particular type of service. For low-temperature and low-pressure service, visual examination may be adequate. Others require some percentage of work be examined by radiographic examination, while the most severe service requires a combination of surface and volumetric examination methods such as visual, magnetic particle, or liquid penetrant examination, and radiographic or ultrasonic examination.

API is the specification for the quality criteria of transmission lines and other petrochemical fabrication and construction. API Standard 650 for Welded Steel Tanks for Oil Storage refers to ASME Section V for Nondestructive Examination methodology, but provides the extent of examination and specific acceptance criteria. API Standard 1104 for Welding Pipelines and Related Facilities provides specific procedure requirements for radiography and acceptance criteria for nondestructive examination results. API has developed a unique, sophisticated ultrasonic examination document designated No. RP-2X for off-shore drilling rigs. This program provides specific techniques and very stringent requirements for personnel qualification. Personnel must demonstrate their capability for ultrasonic examination of T, K, and Y configured weld joints.

NDE Personnel

Nondestructive examination results depend on the performance of personnel. Performance capability of personnel relates directly to qualification, which is usually consistent with the training and experience of the individual. Better training and experience provides the capability to provide consistent, high-quality results. Effective results depend upon well-trained, experienced, and motivated personnel. Personnel assigned to field jobs must have received the training and experience to properly perform the required techniques.

The ASNT Recommended Practice No. SNT-TC-1A: Qualification and Certification for Nondestructive Testing Personnel provides qualification guidelines to the employer for each method.

SNT-TC-1A provides for three basic levels of certification. Level I is qualified to operate equipment and follow written instructions for performance of specific NDE methods. The Level I individual may also evaluate for acceptance or rejection determinations using written instructions and to record the results. Level II individuals are qualified to perform specific nondestructive examinations, interpret results, and prepare NDE reports. A Level III individual is capable of establishing techniques and procedures; interpreting NDE results; interpreting codes and specifications; and training, examining, and certifying other NDE personnel.

Education, training, experience, and examination guidelines are provided for Levels I and II for personnel qualification. The organized training hours are based on education levels. A training outline to cover the body of knowledge is included for each method, and the number of hours varies for each method. The experience requirement is specific to each method, but is consistent regardless of educational background.

Personnel qualification examinations consist of three portions, general, specific, and practical, for Level I and II individuals. The general examination covers the principles and theory of the method. The specific examination covers employer procedures, specific equipment, and applications. The practical examination is a demonstration of the candidate's ability to perform to a given level of qualification with the employer's equipment.

Level III individuals may be certified based on experience as a Level II (or experience in assignments at least comparable to that of a Level II) with employer-based examination. The ASNT NDT Level III certificate with documented evidence of experience, including the preparation of procedures to codes, standards, or specifications, and the evaluation of test results, may be used in lieu of employer's examination for certification. Training for Level III qualification is not currently a specific recommendation in SNT-TC-1A.

ASNT has published a new Standard for the Qualification and Certification of Personnel in Nondestructive Testing (ASNT-CP-189). The new ASNT standard is more stringent than the Recommended Practice No. SNT-TC-1A and provides specific requirements without the flexibility of a recommended practice. This personnel certification standard may be appropriate for some construction NDE to assure high-quality results.

The potential problems associated with unqualified personnel are numerous. Unqualified personnel may let unacceptable materials or workmanship go undetected, resulting in costly failure or even loss of life. Another concern, not usually mentioned, is rejection of acceptable materials. This creates unnecessary costs associated with repair and reexamination. It is imprudent and costly to accept the risk associated with unqualified NDE personnel. If the quality requirements include nondestructive examination, then qualified and certified personnel must be used.

NDE Subcontractors

The construction industry has problems obtaining the level of quality results needed in nondestructive examination. The cost to industry resulting from substandard performance is excessive. Certainly, personnel qualification is essential to good NDE results, but more effort is needed to develop improved and consistent performance of reliable nondestructive examination. Most of the construction industry uses NDE subcontractors to perform the required nondestructive examination. All nondestructive testing agencies are not equal. Some companies use approved vendor lists for the selection of subcontractors. The evaluation of subcontractor programs and personnel is a cost to construction. Where evaluations are not required, they are often eliminated to reduce costs to construction.

A better system is needed to assure more consistent performance of nondestructive examinations. The present cost to industry resulting from substandard performance is excessive. Prequalification of NDE subcontractors' programs and personnel will reduce the costs to construction that are associated with poor performance.

Increased utilization of ASTM Practice for Determining the Qualification of Nondestructive Testing Agencies (E 543) will improve the general qualifications of the nondestructive testing agencies and result in improved performance. This practice establishes minimum requirements for agencies performing nondestructive examination. It is used to assess the capability and abilities of NDT agencies and as a basis for developing an accreditation procedure.

Experience

Recent experience provides several examples that serve to illustrate that the quality of personnel training and certification varies considerably from agency to agency. Some nondestructive testing laboratories do not realize their limitations.

A major mining and metals company had ordered a large semiautogenous (SAG) processing mill. The ductile iron foundry receiving the order used a local nondestructive testing laboratory to perform magnetic particle and ultrasonic examination. The nondestructive testing laboratory accepted the castings to an ASTM product specification which provided the requirements. The client was dismayed to discover during machining that their 30 000 lb (13 608 kg) castings had extremely large areas of dross and had to be scrapped. Recasting the several sections was obviously costly and time consuming.

A surveillance of the subsequent ultrasonic examination of the new castings found ostensibly "qualified" ultrasonic technicians without procedures or the specification for reference. When the specification was reviewed, it was quickly obvious that a required ultrasonic transfer mechanism (adjustment for material attenuation differences) was never performed, which resulted in accepting an unacceptable casting. The nondestructive testing company had no program for maintaining procedures and did not possess the required specifications, but they thought they could perform the examination without these references.

This is what frequently happens when the client is not technically knowledgable concerning nondestructive testing. The client references the quality standards required by the contract. The client expects that the nondestructive testing agency is knowledgable, and they will then receive the quality examination intended. Too often this is not the case and many times is not discovered during construction.

In another case, the ammonia refrigeration system for a food processing plant was modified. The engineers specified the ANSI B31.5 standard to control the fabrication quality. The client was unfamiliar with construction standards and practices. The client did not provide any quality monitoring or overview. They did not realize there could be a problem when they had hired a supposedly "competent" contractor. When one of the welds failed about ten weeks later, thousands of pounds of ammonia spilled into the environment. A small town had to be evacuated due to the intense ammonia vapors and physical danger from the ammonia gas. The plant was forced to discontinue operations for months while the extent of the problem was evaluated and corrected. It should be understood that this particular contractor had been performing work at the facility for about ten years without any obvious problems.

I do not know the origin, but there is an old saying that recognizes this point: "You can expect only what you inspect." Many times in my experience, unexpected nondestructive examination has resulted in extensive repairs to meet the workmanship quality requirements. Experience has demonstrated repeatedly that workmanship quality is significantly greater when nondestructive examination or other quality inspection is planned.

I believe several factors are involved that contribute to this improved performance when testing and inspections are performed. First, management will use their most skilled craft people on work which will be inspected. Secondly, craft people seem to use more care when they know their work will be inspected.

Conclusion

More emphasis must be placed on quality in the construction industry. Constructors are only sensitive to requirements in order to remain competitive. This focus results in a short-term cost and schedule emphasis, sometimes at the expense of workmanship quality. Cost and schedule are the driving forces for this relatively short-term activity because performance is measured in these areas. The cost of this paradigm to industry is far too high to continue. The emphasis must shift to longer term considerations driven by a balance between current costs and schedules and the costs of operating maintenance and unscheduled outages or catastrophic failure.

A combination of changes must come into play. The client must involve operating experience with engineering in developing requirements for construction. Companies need to involve operating personnel in the codes and standards development organizations to influence the changes needed in standard construction practices. In this way the requirements will apply equally to assure the owners that the quality results needed will be achieved. This will result in constructors being more competitive while delivering higher quality products.

Meinhard Stadthaus[1]

Control Methods for Magnetic Particle Inspection

REFERENCE: Stadthaus, M., **"Control Methods for Magnetic Particle Inspection,"** *Nondestructive Testing Standards—Present and Future, ASTM STP 1151,* H. Berger and L. Mordfin, Eds., American Society for Testing and Materials, Philadelphia, 1992, pp. 230–234.

ABSTRACT: Nondestructive inspections are increasingly called upon to follow foreign and international codes. In many cases the user is not familiar with the required inspection procedures and parameter checks. A guideline about procedures and checks for magnetic particle inspection has been published by the German Society of Nondestructive Testing (DGZfP). The following brief version includes additional comments on future developments in standardization.

KEY WORDS: magnetic particle inspection, standardization, specification nondestructive testing

Difficulties sometimes arise in the application of magnetic particle inspection (MPI), as recommended foreign codes are sometimes misunderstood and/or misinterpreted. Because of these difficulties, the German Society of Nondestructive Testing (DGZfP) has published a guideline [*1*] which provides a survey, a comparison, and critical comments on the procedures and parameter checks of important foreign, German, and international codes. This paper reviews the content of the guideline, giving typical examples.

Content

The guideline presents a check of the following inspection parameters:

1. *Inspectability of the material.* Checks of the magnetic ability of the material to guarantee sufficient magnetization for the inspection.
2. *Magnetization.* Values for the magnetic field required for a proper magnetization. Methods to determine the magnetization.
3. *Check of the remanent fields.* Measurement of the remaining magnetic field after demagnetization.
4. *Detection media.* Determination of the quality of the detection media and description and application of the different specimens.
5. *Viewing conditions.* Levels of visible light and the ultraviolet (UV) radiation for nonfluorescent and fluorescent detection media.
6. *Surface conditions.* Cleaning methods and requirements on the surface structure (roughness, nonmagnetic layers).

[1] Federal Institute for Materials Research and Testing (BAM), Unter den Eichen 87, D-1000 Berlin 45, Germany.

7. *Overall performance check.* Check of all inspection parameters on the parts to be inspected with natural or artificial defects.

Each chapter includes:

1. Evaluation of the influence of the parameters on the inspection result, mainly of the visibility of the indications.
2. Description of the methods recommended in the specification involved.
3. Comment on the sense and nonsense of the check as well as the applicability of the results.

Examples

Magnitude of Magnetization

The visibility of an indication depends on the stray flux above the crack (more exact: the stray-flux gradient). This stray flux is proportional to the flux density, B, and the depth of the crack.

The German guideline [2] quotes the value $Bm \geq 1$ T. For low-alloy, low-carbon steels (typical application), this value will be reached at a field strength $H < 1.0$ kA/m on the initial branch of the magnetization curve. Considering a security factor, the minimum value $H = 2.0$ kA/m is established. Values in this region are used in most important international codes [3–5] (conversion: 2 kA/m = 2.5 mT). The value $H = 2.0$ kA/m will probably be specified in the European standard.

Magnetization

Determination of Magnetization—Newer international specifications require a measurement of the tangential field strength, H_t, at least as a basic reference. Considering that H_t on both sides at a boundary (iron/air) is the same, H_i inside the material can be measured outside (see Fig. 1). The field-sensitive area of the probe always has a certain distance, h_0, from the surface. Therefore, a measuring error may occur in the case of inhomogeneous increasing fields. An example is given in Fig. 1 when determining H_t by a second measurement of H in the distance, $2 h_0$, and extrapolating H on the surface. Nevertheless, in grossly inhomogeneous fields this measurement will not be adequate. In Fig. 2 it is shown that the tangential field strength at the end of a cylindrical body in a homogeneous field, H_m, differs strongly from the calculated value H_c and the exact measured value H_ϕ. It should be mentioned that, in the case of inhomogeneous fields and multidirectional magnetization, the determination of sufficient magnetization is more reliable with cracks or spark-eroded slots in the surface. Because of their small thickness and their flexibility, so-called quantitative quality indicators (QQI) or shims can be fixed very close to the surface. Indications of this type depend mainly on the magnetization of the specimen on the surface (not so with the "Berthold" or the "pie gage"). The use of this type of indicator may be very helpful for indicating proper magnetization. Generally, the use of flux shunting indicators is acceptable only as a subjective comparative method. Calculations from the current, as in many U.S. specifications, may be suitable only as a rough estimation of the true magnetization. The reasons are the different geometries and magnetizing conditions which appear in practical applications.

Characterization of Currents—Magnetizing currents with periodic waveforms (e.g., alternating current) as well as the corresponding magnetic fields can be described by characteristics such as the RMS value [1] or the peak value [3]. According to our own investigations [6], the

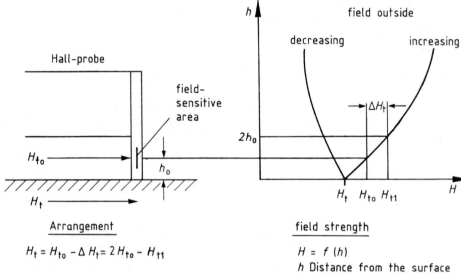

$$H_t = H_{to} - \Delta H_t = 2H_{to} - H_{t1}$$

Arrangement

$H = f(h)$
h Distance from the surface

field strength

FIG. 1—*Determination of the tangential field strength on the surface.*

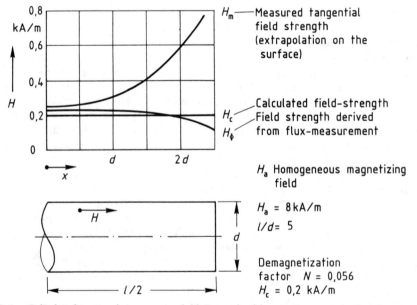

H_m ——Measured tangential field strength (extrapolation on the surface)

H_c ——Calculated field-strength
H_ϕ ——Field strength derived from flux-measurement

H_a Homogeneous magnetizing field

$H_a = 8\,kA/m$
$l/d = 5$

Demagnetization factor $N = 0,056$
$H_c = 0,2\ kA/m$

FIG. 2—*Cylindrical part in a homogeneous field. Example of the measurement error in inhomogeneous magnetized parts.*

visibility of an indication correlates better with the RMS value than with the peak value. The RMS value is commonly used in the field of electrotechnique. This value should be measured and indicated in accordance with new proposals for European standards.

Detection Media

The quality of the detection media shall be evaluated and checked on specimens. The following requirements are valid for an ideal specimen and the flaws therein:

1. The flaws shall have the same characteristics (stray flux topography) as the flaws to be detected. For normal applications these are surface flaws with a width down to 5 μm (0.0002 in.).
2. The flaws shall be produced with well-defined dimensions.
3. The indication above the flaws shall be at the margin of visibility. In this region a good visual valuation of the indications is possible.
4. The specimen should be remanently magnetized (better handling).

These requirements are not fulfilled by any existing specimen. The Ketos ring (holes under the surface) evaluates dry coarse powders better than fine wet powders, which is contradictory to the normal practice for detection of surface defects [6]. The German "MTU Nr. 3" [7] is remanently magnetized and has very small natural cracks but cannot be produced with defined dimensions (ideal for comparative valuation). The French "Specimen C" [8] has the potential to fulfil the above-mentioned requirements. In the corresponding European standard, one "Specimen A" for quantitative valuation ("Specimen C") and one "Specimen B" (MTU Nr. 3) for normal comparative applications may be established.

Overall Performance Check

It is evident that the best way of checking magnetic particle inspection is the application of the method on parts with typical cracks with the smallest dimensions to be detected in accordance with the specification. But this "overall performance check" is only possible if a sufficient number of parts with such cracks are available (possible only in mass production). Another possibility is the incorporation of spark-eroded slots with a minimum width of approximately 50 μm (0.002 in.). Thin and flexible flux shunting indicators such as the QQI can be attached close to the surface (see *Magnetization*). EDM slots, as well as the QQI, do not fulfill the requirements for a check of the detection media (see *Detection Media*), but are especially useful for checking magnetization. The check of the detection medium must be verified separately.

Conclusions

This brief description of the guideline as well as the additional remarks point out some problems with magnetic particle inspection. The guideline (finished in 1989) may be helpful in designing new types of specifications which should be harmonized worldwide. Until now international standardization (ISO) has not been as successful as it should be. The activities of CEN (European standardization) have increased because standards should be available to the common market in 1993. In the field of magnetic particle inspection, the working group TC 138/WG5 began with "Terminology" and "General Rules." In addition, "Magnetizing Equipment" and "Detection Media" are in preparation. The working group, Magnetic Particle

Inspection of Welds (TC 121/WG5B/SWG5), is publishing a proposal, "General Rules for Magnetic Particle Inspection of Welds," and is working on "Acceptance Criteria" (introduction of acceptance levels and quality classes). A proposal, "Magnetic Particle Inspection of Forgings," which includes quality classes, is in preparation (ECISS/TC 28/WG1/SG1). It is evident that the harmonization of these standards will be difficult because special codes (e.g., product codes) and important international specifications must be taken into consideration.

References

[1] DGZfP-EM3 Comments on Checking Inspection Parameters in Magnetic-Particle-Inspection, 1991, English version.

[2] DGZfP-EMO Guideline on Magnetic-Particle Inspection, 1987, English version.

[3] BS 6072 (British standard) Method for Magnetic-Particle-Flaw-Detection, 1981.

[4] NF A 09–125 (AFNOR) Principes Generaux de L'examen Magnetoscopique, 1982.

[5] MIL STD 1949 Inspection, Magnetic Particle, 1985.

[6] Stadthaus, M., Dickhaut, E., Prestel, E., "System Performance-Control in Magnetic-Particle-Inspection," TT2/13, *Proceedings of the Fourth European Conference on Nondestructive Testing*, London, 1987.

[7] DBP 2357220 (German patent) Verfahren zur Herstellung eines metallichen Prufkorpers zur Uberwachung von Magnetpulver-Prufflussigkeiten.

[8] A 09-570 (AFNOR) Essais non destructifs magnetoscopie. Caracterisation des produits, 1988.

Author Index

Subject Index

A

Acceptance criteria for NDT standards, 41

Acoustic emission standards, 56–62

Acousto-ultrasound, 56

AE standards, 56–62

Aerospace applications—high-temperature materials, NDE standards, 211

Aerospace council technical organization, 156

Aerospace industry, SAE/AMS NDT standards, 153

Aerospace materials specifications, 153–162

American National Standards Institute(ANSI), data management of NDT standards, 203–210
standards
  B31.1, 226
  B31.3, 226
  B31.5, 228

American Petroleum Institute(API)
standards
  API 650, 226
  API 1104, 226
  RP-2X, 226

American Society of Mechanical Engineers (ASME), 136, 185
ASME code
  boiler and pressure vessel code, 136
  legal status, 136
  nondestructive testing, 136–152
  personnel qualification, 185–194

American Society of Nondestructive Testing (ASNT), 136
recommended practice
  CP-189, 227
  SNT-TC-1A, 226–227

American Welding Society(AWS), 226

ANSI. *See* American National Standards Institute.

ASNT. *See* American Society of Nondestructive Testing.

ASTM program for nondestructive testing standards
committee E-7, 9–14
  administrative subcommittees, 10(table)
  technical subcommittees, 11(table)
overview, 1

ASTM standards
Committee E-7 overview, 9–14
list of ASTM standards for international development, 165(table)
standards
  E-94, 166, 203
  E-125, 42
  E-165, 42, 165
  E-433, 42
  E-543, 75, 225, 228
  E-545, 167
  E-569, 166
  E-610, 166
  E-709, 42
  E-746, 167
  E-747, 167
  E-748, 167
  E-999, 203
  E-1001, 166
  E-1025, 167
  E-1030, 166
  E-1032, 166
  E-1106, 166
  E-1133, 42
  E-1135, 42
  E-1208, 42
  E-1209, 42
  E-1210, 42
  E-1212, 75
  E-1220, 42
  E-1254, 167, 203
  E-1316, 42
  E-1324, 166
  E-1359, 75
  E-1391, 75
  E-1417, 42
  E-1418, 42
  E-1444, 42